T0073339

MEMS Silicon Oscillating Accelerometers and Readout Circuits

RIVER PUBLISHERS SERIES IN CIRCUITS AND SYSTEMS

Series Editors:

MASSIMO ALIOTO
National University of Singapore
Singapore

KOFI MAKINWA
Delft University of Technology
The Netherlands

DENNIS SYLVESTER
University of Michigan
USA

Indexing: All books published in this series are submitted to the Web of Science Book Citation Index (BkCI), to SCOPUS, to CrossRef and to Google Scholar for evaluation and indexing.

The "River Publishers Series in Circuits & Systems" is a series of comprehensive academic and professional books which focus on theory and applications of Circuit and Systems. This includes analog and digital integrated circuits, memory technologies, system-on-chip and processor design. The series also includes books on electronic design automation and design methodology, as well as computer aided design tools.

Books published in the series include research monographs, edited volumes, handbooks and textbooks. The books provide professionals, researchers, educators, and advanced students in the field with an invaluable insight into the latest research and developments.

Topics covered in the series include, but are by no means restricted to the following:

- Analog Integrated Circuits
- Digital Integrated Circuits
- Data Converters
- Processor Architecures
- System-on-Chip
- Memory Design
- Electronic Design Automation

For a list of other books in this series, visit www.riverpublishers.com

MEMS Silicon Oscillating Accelerometers and Readout Circuits

Editor

Yong Ping Xu

National University of Singapore
Singapore

Routledge
Taylor & Francis Group

LONDON AND NEW YORK

Published 2019 by River Publishers
River Publishers
Alsbjergvej 10, 9260 Gistrup, Denmark
www.riverpublishers.com

Distributed exclusively by Routledge
4 Park Square, Milton Park, Abingdon, Oxon OX14 4RN
605 Third Avenue, New York, NY 10017, USA

First issued in paperback 2023

MEMS Silicon Oscillating Accelerometers and Readout Circuits / by Yong Ping Xu.

Routledge is an imprint of the Taylor & Francis Group, an informa business

Publisher's Note
The publisher has gone to great lengths to ensure the quality of this reprint but points out that some imperfections in the original copies may be apparent.

While every effort is made to provide dependable information, the publisher, authors, and editors cannot be held responsible for any errors or omissions.

ISBN 13: 978-87-7022-966-1 (pbk)
ISBN 13: 978-87-7022-045-3 (hbk)
ISBN 13: 978-1-003-33882-6 (ebk)

Contents

4   An MEM Silicon Oscillating Accelerometer Employing a PLL and a Noise Shaping Frequency-to-Digital Converter   93

Jian Zhao, Yong Ping Xu and Yan Su

Preface

The advances of Microelectromechnical systems (MEMS) and integrated circuit (IC) technologies have steadily improved the performance of MEMS inertial sensors, namely MEMS accelerometers and gyroscopes. MEMS accelerometer with Application Specific IC (ASIC) can be found in many applications, such as gaming, mobile devices, robots, industrial control and inertial guidance & navigation, for its good performance, small form factor, low cost and low power consumption.

The sensing mechanisms of mainstream MEMS accelerometers are based on the well-known Hooke's law and Newton's laws. The sensing element is a proof mass supported by a spring attached to a reference frame. Based on the Hooke's law, the force and hence the acceleration can be sensed by the displacement of the proof mass, referring to as "Displacement sensing mechanism". This gives rise to the "capacitive MEMS accelerometers", since the displacement of the proof mass can be detected by the capacitance change of the electrodes attached to it. Most commercial accelerometers on the market belong to this category. MEMS silicon oscillating accelerometer (SOA) or resonant accelerometer, on the other hand, detects the acceleration via the inertial force applied to the proof mass, where the force is detected by a change in the resonant frequency of the MEMS resonator whose beam is loaded by the proof mass. This is referred to as "Force sensing mechanism". To achieve high sensitivity and signal-to-noise ratio (SNR), MEMS SOA is required to operate in a vacuum environment (vacuum packaged) in order to maintain a high quality factor (high-Q) for the MEMS resonator.

The first MEMS SOA with an integrated readout circuit was reported by Roessig from UC Berkeley [1] in 1997. It was a single chip SOA with on-chip readout circuit and fabricated in a BiMEMS surface-micromachining process. The accelerometer achieved a scale factor of 2.4 Hz/g and bias instability of 15.8 mg with a full scale of ±20 g. An improved design based on Sandia Laboratories' MEMS/CMOS surface micromachining process achieved 45 Hz/g scale factor and 89 μg bias instability [2]. In 2002, Seshia from UC Berkeley reported an improved design from Roessig's early work. It was based on the same surface micromachining technology from Sandia National Laboratories with the integrated readout circuit and the accelerometer achieved a scale factor of 17 Hz/g and a noise floor of 40 $\mu g/\sqrt{Hz}$. Later in 2008, another MEMS SOA with CMOS readout ASIC was reported by He from National University of Singapore [3]. In this work, the MEMS sensor and ASIC were fabricated in a 20-μm Silicon-on-Insulator (SOI) process and 0.35-μm CMOS, respectively, and it achieved 4 μg bias instability and 20 $\mu g/\sqrt{Hz}$ noise floor with a scale factor of 140 Hz/g and a power consumption of only 23 mW. MEMS SOAs with board-level readout circuits were also reported. Among them, Draper Labs demonstrated a MEMS SOA prototype in 2000, which was fabricated in a 12-micro-m Silicon-on-Glass (SOG) process [4]. It achieved a low white noise floor at 23 $\mu g/\sqrt{Hz}$, bias stability of 5 μg and sub-μg bias instability. Later in 2005, Draper Labs reported another two MEMS SOAs with high-g and lower-g full ranges, respectively [5]. These two prototypes demonstrated superior performances with 0.5 and 0.08 μg for bias instability, 0.92 and 0.19 μg for bias stability, and 0.56 and 0.14 ppm for scale factor drift, respectively. These early research outcomes had shown that the MEMS SOA could be a good candidate for high-performance MEMS accelerometers.

This book covers recent developments of MEMS SOAs. It consists of six chapters and includes both designs of MEMS sensors and readout circuits, as well as the modeling of the SOA system. The book is therefore useful for researchers, engineers and postgraduate students, who work on either mechanical design of the MEMS sensor or the ASIC design.

Chapter 1 describes mechanical design of the MEMS sensor in the SOA. It starts with a brief introduction of MEMS SOA and the theory of operation and then discusses the design of MEMS sensor with a focus on the micro-lever design. Performance metrics, including scale factor, bias and thermal sensitivity of the MEMS SOA, are derived and the effect of stiffness nonlinearity is discussed. It ends with a brief introduction of the fabrication process and the measured performance parameters of several fabricated MEMS sensor prototypes. Chapter 2 deals with front-end amplifiers in the MEMS SOA, which serves as the interface between the MEMS sensor and readout circuit. Various front-end amplifier designs for both MEMS oscillators and SOAs are reviewed and discussed since the core of the SOA readout circuit is essentially an MEMS oscillator. Chapter 3 focuses on the readout circuit design for a MEMS SOA. It discusses the key design considerations and analyzes the phase noise and amplitude-stiffness effect. A readout circuit fabricated in a 0.35-μm CMOS process is presented with the measured performance of the complete SOA. Chapter 4 presents a MEMS SOA with a different architecture. In this new architecture, the drive signal amplitude is set by a low noise reference voltage, while the phase of the drive signal is provided by a PLL that is locked to the phase of the MEMS oscillator output. A complete MEMS SOA system is presented with the measured performance. Chapter 5 introduces a system-decomposition phase noise model for MEMS SOAs. The model proposes a unified approach to decompose the physical system of the MEMS SOA into phase and amplitude subsystems to predict the phase noise of MEMS oscillators and the performance of the MEMS SOA. The model is tested with both analytical model in MATLAB and time-domain simulation in SIMULINK. Chapter 6 presents a MEMS SOA aimed at monitoring seismic activity, which requires the accelerometer to have an ultra-low noise floor. It describes the design of the MEMS sensor in detail. The drift of sensor output and its sources are analyzed. MEMS sensor fabrication and calibration are presented in the end with some measured results.

The book would not have been possible without the contributions and support from the following people. First, I would like to acknowledge the contributions from chapter authors and co-authors. They are the authors of Chapter 1, Professor Anping Qiu from Nanjing University of Science and Technology, China, and Chapter 6, Professor Xudong Zou from Institute of Electronics, Chinese Academy of Science, China, and co-authors of Chapter 2–5, Professor Yan Su and Dr Yang Zhao from Nanjing University of Science and Technology, Dr Jian Zhao from Tsinghua University, China, and Dr Xi Wang from China University of Mining and Technology, China. I would also like to thank Mr Mark de Jough from River Publisher, who initially approached me for this book, for his help and patience throughout the preparation process. Last but not least, Production manager, Ms Junko Nakajima, and her staff have provided the timely production of the book.

Yong Ping Xu

References

[1] T. A. Roessig, R. T. Howe, A. P. Pisano and J. H. Smith, "Surface-micromachined Resonant Accelerometer," in *Transducers'97*, pp. 859–862, 1997.

[2] T. Roessig, "Integrated MEMS tuning fork oscillators for sensor applications," Ph.D. dissertation, Department of Mechanical Engineering, University of California, Berkeley, 1998.

[3] L. He, Y. P. Xu and M. Palaniapan, "A CMOS Readout Circuit for SOI Resonant Accelerometer With 4-µg Bias Stability and 20-µg/$\sqrt{\text{Hz}}$ Resolution," *IEEE Journal of Solid-State Circuits*, vol. 43, no. 6, pp. 1480–1490, 2008.

[4] R. Hopkins et al., "The silicon oscillating accelerometer: A MEMS inertial instrument for stratergic missle," in *Guidance Missile Sci. Conf.*, pp. 44–50, 2000.

[5] R. Hopkins et al., "The Silicon Oscillating Accelerometer: High-Performance MEMS Accelerometers for Precision Navigation and Stratergic Guidance Applications," in *The 2005 National Technical Meeting of The Institute of Navigation (ION)*, pp. 971–979, 2005.

List of Contributors

Anping Qiu, *No. 200 Xiaolingwei, Xuanwu District, Nanjing, China;*
E-mail: apqiu@njust.edu.cn

Yan Su, *Nanjing University of Science and Technology, Nanjing, China;*
E-mail: suyan@mail.njust.edu.cn

Xi Wang, *School of Information and Control Engineering, China University*
of Mining and Technology, Jiangsu, China;
E-mail: icwangxi@cumt.edu.cn

Yong Ping Xu, *Department of Electrical & Computer Engineering, National*
University of Singapore, Singapore; E-mail: yongping@ieee.org

Jian Zhao, *Tsinghua University, Beijing, China;*
E-mail: zhaojianycc@mail.tsinghua.edu.cn

Yang Zhao, *Nanjing University of Science & Technology, Nanjing, China;*
E-mail: zhaoyang0216@njust.edu.cn

Xudong Zou, *State Key Laboratory of Transducer Technology, Institute of*
Electronics, Chinese Academy of Sciences, Beijing, China;
E-mail: xdzou@mail.ie.ac.cn

List of Figures

List of Tables

List of Abbreviations

3-D	Three-Dimensional
AAC	Auto amplitude control
AC	Alternating Current
ADC	Analog-to-digital converter
AGC	automatic gain control
AM	Amplitude modulation
AS	amplitude-stiffness
A-S	Amplitude-stiffening
ASIC	application-specific integrated circuit
AVAR	Allan Variance
AVF	alternating voltage follower
AXL	Accelerometer
BOX	Buried Silicon Oxide
BW	bandwidth
C-C Beam	Clamped-Clamped Beam
CDS	correlated double sampling
CMFB	common-mode feedback
CMOS	complementary metal oxide semiconductor
CSA	Charge-sensing amplifier
CT	continuous time
CTE	Coefficient of Thermal Expansion
D2S	Differential-to-single
DAQ	Data acquisition
DC	Direct Current
DCO	Digital controlled oscillator
DETF	double ended tuning fork
DFF	D-flip flop
DRIE	Deep Reaction Ion Etch
DT	discrete time
DZ	Dead zone
FDC	Frequency-to-digital converter

FEA	Finite Element Analysis
FEM	Finite Element Method
FM	Frequency Modulation
GBW	gain-bandwidth product
GT	Gate Time
ICP	inductively-coupled plasma
ISF	impulse sensitive function
I-V	current to voltage
LCC-44	Leadless Chip Carrier with 44 pins
LCR Circuit	An electrical circuit consisting of, an inductor, a capacitor and a resistor
LTI	Linear-time invariant
LTV	Linear-time variant
MDEV	Modified Allan Deviation
MEMS	Micro-electro-mechanical system
MPLL	Modified phase-locked loop
MVAR	Modified Allan Variation
NRF	Noise transfer function
Opamp	operational amplifier
OTA	Operational transconductance amplifier
PCB	Printed circuit board
PD	Phase detector
PFD	Phase and frequency detector
PLL	Phase-locked loop
PM	Phase modulation
PSD	Power spectrum density
PSG	Phosphosilicate Glass
RIE	Reaction Ion Etch
RMS	Root Mean Square
RXL	Resonant Accelerometer
SDM	Sigma-delta modulator
SNDR	Signal-to-noise and distortion ratio
SNR	signal-to-noise ratio
SOA	Silicon oscillating accelerometer
SOI	Silicon On Insulator
STF	Signal transfer function
TCE	Temperature Coefficient of Elasticity
TCf	Temperature Coefficient of the output frequency of the MEMS RXL

TCS	Temperature Coefficient of the scale factor of the MEMS RXL
TIA	Trans-impedance amplifier
VBA	vibration beam accelerometer
VCO	Voltage controlled oscillator
VGA	Variable gain amplifier
VRW	velocity random walk

List of Notations

c_0	damping coefficient of the MEMS resonator
e	strain along the resonant beam
g_m	transconductance
h_n	the power density of f^n phase noise at 1 Hz
h_{add}	impulse response of the MEMS resonator
$h_{add}^{\langle p \rangle}$	additive phase-mode impulse response of the MEMS resonator
$h_{add}^{\langle a \rangle}$	additive amplitude-mode impulse response of the MEMS resonator
$h_{mul}^{\langle p \rangle}$	multiplicative phase-mode impulse response of the MEMS resonator
$h_{mul}^{\langle a \rangle}$	multiplicative amplitude-mode impulse response of the MEMS resonator
$h_r^{\langle p \rangle}$	phase-mode impulse response of the MEMS resonator
$h_r^{\langle a \rangle}$	amplitude-mode impulse response of the MEMS resonator
k_0	linear stiffness of the MEMS resonator
k_2	cubic nonlinear stiffness of the MEMS resonator
k_{eff}	effective stiffness of the MEMS resonator
k_{vp}	a constant dependent on the dimension of variable-gap capacitors in the MEMS electrodes
k_B	Boltzmann's constant
m_0	lumped mass of the resonator beam
n_{add}	additive noise
n_{mul}	multiplicative noise
$n^{\langle p \rangle}$	the decomposed phase-mode noise
$n^{\langle a \rangle}$	the decomposed amplitude-mode noise
x_0	displacement amplitude of the resonant beam
x_{add}	additive response of the displacement amplitude

x_{mul}	multiplicative response of the displacement amplitude
x_r	the total amplitude response
α_i	the PSD of the white noise on node i
β_i	the PSD of the flicker noise on node i
γ	the constant related to the operation regions of the MOSFET
δ	impulse function
δ_{mul}	multiplicative impulse function
ϵ	normalized position along the resonant beam
ϵ_0	the permittivity of vacuum
ϕ_i	the i-th order mode shape function of the resonant beam
θ_r	the total phase response
θ_{add}	additive phase response
θ_{mul}	multiplicative phase response
ω_0'	resonant frequencies with nonlinear stiffness
ω_0	resonant frequencies without nonlinear stiffness
ω_c	cut-off frequency of the decomposed transfer function
A_0	cross-section area of the resonant beam
E	elastic modulus of the resonant beam
$F_{thermal}$	thermal noise equivalent force
$F_{elastic}$	elastic force of MEMS resonator
$H_r^{\langle p \rangle}$	decomposed phase-mode transfer function
$H_r^{\langle a \rangle}$	decomposed amplitude-mode transfer function
K_m	mechanical gain of the resonant beam
L_0	length of the resonant beam
M	modulation matrix
Q	quality factor
R	the equivalent motional resistance of a mechanical system
S_i	the PSD of the input noise on node i
S_{unit}	the PSD of the unit noise source in the simulation
S_ψ	power spectrum density of the frequency noise
S_ϕ	power spectrum density of the phase noise
T	absolute temperature
V_b	bias voltage for the MEMS resonator
$\overline{V_n^2}$	the PSD of the white thermal noise in MOSFET
$\overline{V_{fn}^2}$	the PSD of the flicker noise in MOSFET
f_{in}	frequency of the input signal

f_{osc}	frequency of the oscillator
f_{BW}	bandwidth of the output signal
f_{vco}	center frequency of the VCO
k_d	gain of the phase detector in the linearized MPLL-FDC
k_{vco}	gain of the VCO
m_d	equivalent lumped mass of the resonator beam in drive mode
q_n	time domain quantization noise
x_0	displacement amplitude of the resonant beam
$\Delta\theta$	phase difference between two input signal of the PFD
Φ_{CLK}	phase of the clock
Φ_q	quantization phase error
γ	gain ratio between analog multiplier and DZ-PFD in the hybrid PFD
ω_0	resonant frequencies without nonlinear stiffness
ω_p	frequency of the pole in the loop filter
ω_z	frequency of the zero in the loop filter
C	digital output of the counter
D_{out}	digital output of the FDC
F_{out}	measured frequency output of the FDC
G_L	overall loop gain of the PLL
K_m	mechanical gain of the resonant beam
K_{CP}	gain of the charge pump in the hybrid PFD
K_{mul}	gain of the analog multiplier in the hybrid PFD
K_{cmp}	gain of the comparator
K_{bd}	the mechanical gain from drive force to displacement @ ω_0
N_{ref}	noise in the voltage reference V_{ref}
N_q	quantization noise in s domain
Q_d	quality factor of the resonator beam in drive mode
T_{in}	period of the input signal
T_{CLK}	period of the clock signal
V_{ref}	reference voltage for amplitude control
V_b	bias voltage of the MEMS resonator beam
V_{op}	output voltage of the comparator
A_{ad}	gain of the amplitude detector
A_{TIA}	pass-band gain of the TIA
$I_{n,TIA}$	input-referred current noise spectrum density of the TIA

I_{sense}	sensing current of the resonator
$K_{X/I}$	gain from displacement to current of the resonator
$N_{1/f,AAC}$	lumped 1/f noise in AAC circuit at input of error amplifier
$N_{1/f,BIAS}$	1/f noise in bias voltage to resonator
$N_{1/f,REF}$	1/f noise in V_{ref}
$N_{1/f,VGA}$	1/f noise referred to the DC input node of VGA
V_{ref}	input reference voltage to control oscillation amplitude
V_p	polarization voltage for MEMS resonant beam
C_{s0}	static capacitance of sense comb electrode
n_s	number of comb fingers
L_0	overlap length of sense comb
g_0	gap of sense comb
h	thickness of sense comb
x(t)	oscillating displacement of MEMS resonant beam
$V_d(t)$	drive signal for MEMS resonator
ω_0	oscillating frequency of MEMS resonator
R_f	feedback resistor in resistive front-end
C_c	stability compensation capacitor in resistive front-end
C_f	feedback capacitor in capacitive front-end
R_b	DC feedback resistor in capacitive front-end
C_{in}	input loading capacitor of front-end
I_s	motion current of MEMS resonator
A_0	open loop DC gain of OTA/Opamp
ω_{p1}	dominant pole of OTA/Opamp
R_{eq}	equivalent overall transimpedance of front-end
k	Boltzmann constant
T	temperature
R_m	equivalent motion resistance of MEMS resonator
C_m	equivalent motion capacitance of MEMS resonator
L_m	equivalent motion inductance of MEMS resonator
Q	quality factor of MEMS resonator
K_{vf}	capacitive transducer gain of drive comb
K_{vi}	capacitive transducer gain of sense comb
V_c	output of automatic amplitude control loop
V_{ni}	input referred voltage noise of OTA/Op-amp
I_n	current noise of feedback resistor
f_s	sampling frequency

A_D	the cross-sectional area of the DETF resonator
B	the bias of SOA
B_T	the sensitivity of the bias to temperature
E	Young's modulus of the material
f_0	the natural frequency of the DETF resonator without axial force
f_n	the natural frequency of the DETF resonator
$\Delta\phi$	the difference of the natural frequencies of the DETF resonators
F	the axial load applied to the beam of the DETF resonator
F_0	an initial axial force acting on the beam of the DETF resonator
F_{in}	input force
F_{ho}, F_{hp}, F_{hi}	the tangential force at the output beam, the pivot beam and the input beam, respectively
h	the depth of the beam of the DETF resonator
I	moment of inertia of cross section of the DETF resonator
$k_{hmo}, k_{hmp}, k_{hmi}$	the horizontal displacement of the output beam, pivot beam and input beam under the bending moments, respectively
$k_{hho}, k_{hhp}, k_{hhi}$	the tangential stiffness of the output beam, pivot beam and input beam under the tangential forces, respectively
$k_{\theta mi}, k_{\theta hi}$	the bending stiffness of the input beam when loaded with M_i and under the tangential force F_{hi}, respectively
$k_{\theta mo}, k_{\theta ho}$	the bending stiffness of the output beam when loaded with M_o and under the tangential force F_{ho}, respectively
$k_{\theta mp}, k_{\theta hp}$	the bending stiffness of the pivot when loaded with M_p and under the tangential force F_h, respectively
K_0	the bending stiffness of the beam of the DETF resonator without axial force
K_1	the scale factor of SOA
K_3	the nonlinearity stiffness
K_a	the elastic constraint stiffness of the beam of the DETF resonator

K_{in}	the equivalent axial stiffness
K_m	the bending stiffness of the input axis of SOA
K_{eff}	the effective bending stiffness of the beam of the DETF resonator
l	the length between the input beam and output beam
L	the length between the pivot beam and input beam
L_D	the length of beam of the DETF resonator
m	the central mass of the beam of the DETF resonator
M	the proof mass of SOA
M_{eff}	the effective mass of the beam of the DETF resonator
M_o, M_p, M_i	the bending moments at the output beam, the pivot beam and the input beam respectively
n	the amplification factor of micro lever
n'	the system amplification factor
p_i	constant value related to the i-th shape
P	electrostatic force exerted on comb drive actuator, which is equivalent to a concentrated load
P_v	electrostatic force exerted on comb drive actuator
P_{eff}	the effective electrostatic force, which is equivalent to a concentrated load
q_i	modal coordinate of the i-th shape of the beam of the DETF resonator
u	the axial strain along the beam axis of the two DETF resonator
w	the deflection of the beam of the DETF resonator
w_D	the width of the beam of the DETF resonator
$\Delta\omega_\Delta$	the width mismatch between the two DETF resonators
x_a	the displacement of the proof mass caused by input acceleration
$\gamma_{mo}, \gamma_{mp}, \gamma_{mi}$	the horizontal displacement of the output beam, the pivot beam and the input beam under the bending moments, respectively
$\gamma_{ho}, \gamma_{hp}, \gamma_{hi}$	the horizontal displacement of the output beam, the pivot beam and the input beam under the tangential forces, respectively.
δ	the vertical displacement at the pivot movable end
ε	the ratio of x and L_D
η	the nonlinearity of the scale factor
θ	the rotation angle of the pivot under loading

θ_{mi}, θ_{hi}	the rotation angle of the input beam caused by the bending moment M_i and the tangential force F_{hi}
θ_{mo}, θ_{ho}	the rotation angle of the output beam caused by the bending moment M_p and the tangential force F_{hp}
θ_{mp}, θ_{hp}	the rotation angle of the pivot caused by the bending moment M_p and the tangential force F_{hp}
ρ	density of the material
σ_2	the applied thermal stress on the beams of the resonators
ϕ_ι	ith mode shape of the beam of the DETF resonator
ω_ν	the circular natural frequency of the DETF resonator
$\alpha(t)$	the amplitude fluctuation term of a signal
$\alpha_{Si}^{(1)}$	the first order thermal expansion coefficient of silicon
$\alpha_{Si}^{(2)}$	the second order thermal expansion coefficient of silicon
$\alpha_{Si}^{(3)}$	the third order thermal expansion coefficient of silicon
α_{T_DC}	the temperature coefficient of DC bias voltage
$\beta(j\omega)$	the feedback transfer function of the oscillator circuit
$\Delta\omega'_{1\_AC}$	the angular frequency shift of the DETF fundamental mode induced by the magnitude change of AC driving voltage
$\Delta\omega'_{1\_DC}$	the angular frequency shift of the DETF fundamental mode induced by the change of DC bias voltage
Δf	the resonant frequency change of DETF sensing element
Δf_n	the noise-limited frequency fluctuation of the oscillator output signal
Δf_{out}	the output frequency change of MEMS RXLs
ΔT	the change of temperature
$\Delta V_P(t)$	the fluctuation of DC bias voltage
$|\Delta V_d(t)|$	the voltage magnitude fluctuation of AC driving signal
$\Delta(x)$	Dirac delta function of x
Δx	the displacement on x axis
δy	the displacement on y axis
ϵ	Permittivity
θ	the rotary angle of lever beam of micro-lever
$\theta(t)$	the phase modulation term of a signal

φ	the phase of an ideal signal
$\varphi(t)$	the phase fluctuation term of a signal
η	the nonlinear damping coefficient of DETF
λ	the linear damping ratio of DETF
μ_{air}	the viscosity coefficient of air
ν_0	same as f_0 (Section 3)
$\overline{\nu_0}$	same as $\overline{f_0}$ (Section 3)
ρ_{Si}	the density of silicon
$\rho_{Si,\Delta T}$	the density of silicon after considering thermal expansion
σ	the squeeze number of the squeeze film air damping
$\sigma_y(\tau)$	the Allan variance for the average time t
τ	the relaxation time of the DETF sensing element (Section 3.3)
	the average time of Allan Variance calculation
τ_d	the temporal scale limit of the frequency drift impact on oscillator output signal
$\phi_i(x)$	the i th mode-shape function of C-C beam
$\Psi_H(t)$	the fast time-varying phase fluctuation
$\Psi_L(t)$	the slow time-varying phase fluctuation
ω	the angular frequency of AC signal
ω_0	the angular frequency of an ideal noise/drift free signal
ω_1	the angular frequency of the DETF fundamental resonant mode
ω_1'	the angular frequency of the DETF fundamental resonant mode after considered mechanical and electrostatic nonlinear effects
ω_{1e}'	the resonant angular frequency of DETF modulated by DC bias voltage and the magnitude of AC driving signal
ω_c	the resonant angular frequency of the fundamental anti-symmetric mode of DETF sensing element
ω_i	the angular frequency of the i th mode of DETF sensing element
A	the ideal gain of amplifier
A_{Ele}	the planar area of attached electrode on DETF tine beam
$A_{Ele,\Delta T}$	the planar area of thermal expanded attached electrode on DETF tine beam

A_{Ideal}	the ideal force amplification factor of micro-lever
A_{Lvr}	the force amplification factor of micro-lever
a_1	the maximum vibration amplitude of DETF in the fundamental resonant mode
	the linear voltage gain of an amplifier (Section 3.3.2)
a_2	the 2nd order factor of voltage gain of an amplifier
$a_{\Delta T,E}$	the equivalent acceleration output drift of MEMS RXL induced by the temperature dependent Young's Modulus of silicon
a_c	the nonlinear limit of the DETF vibration amplitude
$\{a_i\}$	the raw temporal sequence of acceleration measurement of MEMS RXL
$\{a_i'\}$	the filtered temporal sequence of acceleration measurement of MEMS RXL
a_{in}	input acceleration signal
$a_{in}(t)$	the measured input acceleration data of MEMS RXL (Section 5)
$a_{in,Max}$	maximum input acceleration limited by the linearity of MEMS RXL
$a_{load,Max}$	maximum input acceleration limited by the robustness of MEMS RXL
a_{min}	the minimum detectable acceleration signal/The resolution of MEMS RXL
$a_{min}(\tau)$	the resolution of MEMS RXL estimated from experimentally measured Modified Allan variance (Section 5) with different average time t
a_{min_amp}	the amplifier noise limited resolution of MEMS RXL
a_{min_DC}	the DC bias voltage noise limited resolution of MEMS RXL
$a_{min\_thermal}$	the mechanical-thermal noise limited resolution of MEMS RXL
B	the measurement bandwidth of MEMS RXL
BW_{Mech}	mechanical bandwidth of MEMS RXL
$b_i f^i$	the i th power-law term of the phase noise processes
C	Capacitance
$C_{0,d}$	the static capacitance between the drive electrode and DETF
$C_{0,s}$	the static capacitance between the sense electrode and DETF

$C_{comb}(x)$	the function of capacitance of comb drive transducers
C_{eq}	the equivalent capacitance of the linear electrical circuit model of DETF
C'_{eq}	the equivalent capacitance of the electrical circuit model of DETF included mechanical and electrostatic nonlinear effects
C_{eq3}	the equivalent capacitance for the cubic electrostatic spring coefficient in the electrical circuit model of DETF
C_{ft}	the feed-through capacitance between drive and sense electrodes of DETF sensing element
$C_{pal}(x)$	the function of capacitance of parallel-plate transducers
$C_{p,d}$	the parasitic capacitance between the drive electrode and the substrate
$C_{p,s}$	the parasitic capacitance between the sense electrode and the substrate
c_0	the damping coefficient of DETF in vacuum
c_{vd}	the squeeze film air damping coefficient of DETF
D_1	the distance between pivot beam and connection beam of micro-lever
D_2	the distance between pivot beam and input beam of micro-lever
d_T	the deflection of DETF tine beam
E_0	the Young's Modulus at reference temperature
E_c	the energy stored in a vibrating beam at its resonant frequency
$E_{sc,si}$	the Young's Modulus (second-order tensor) of single crystalline silicon
EA_{Lvr}	the effective amplification factor of micro-lever
$e(t)$	the fluctuation of the voltage gain of an amplifier
F_{Act}	the transverse actuation force applied on DETF tine beam
$|F_{Act}|_c$	the nonlinear limit of the actuation force applied on DETF
F_{Axial}	the axial force applied on DETF sensing element
F_{Lvr_in}	the input force to micro-lever
F_{Lvr_out}	the output force from micro-lever
F_x	the force applied on x axis

F_y	the force applied on y axis
F_z	the force applied on z axis
f	the Fourier frequency
$f(t)$	the measured output frequency data of MEMS RXL (Section 5)
f_0	the frequency of ideal oscillator output signal
	the natural output frequency of MEMS RXL (Section 5)
$\bar{f}_0$	the average frequency of oscillator output signal over certain period
f_c	the resonant frequency of the fundamental anti-symmetric mode of DETF sensing element
$f_{c,damp}$	the resonant frequency of the fundamental anti-symmetric mode of DETF sensing element after considering the squeeze film air damping
$f_{c,\alpha}$	the resonant frequency of the fundamental anti-symmetric mode of DETF sensing element considered the impact of thermal expansion
f_L	the Leeson Frequency
f_{T_DC}	the resonant frequency of DETF sensing element changed by the temperature drift of DC bias voltage
$G_{(j\omega)}$	the close-loop transfer function of oscillator circuit
G_m	the gain of sustaining amplifier in oscillator circuit
g_{Elc}	the transduction gap between two adjacent electrodes of transducers
$g_{Elc,\Delta T}$	the transduction gap between two adjacent electrodes of transducers considered the impact of thermal expansion
I_{dc}	DC current
I_{shot_noise}	the electronic shot noise current
$I_{z,h}$	the second moment of cross area of horizontal segment of snake beam suspensions
$I_{z,v}$	the second moment of cross area of vertical segment of snake beam suspensions
i_m	the motional current of DETF sensing element
i_s	the current measured from sensing electrodes
$J_i(x)$	the i th order Bessel Function of x
K_{eff}	the effective stiffness of DETF tine beam

K_I	the electromechanical coupling coefficient of the sensing transducer
K_P	the electromechanical coupling coefficient of the driving transducer
$k_{\theta c}$	the rotational stiffness of connection beam of micro-lever
$k_{\theta p}$	the rotational stiffness of pivot beam of micro-lever
k_{Lvr_in}	the equivalent input stiffness of micro-lever
k_{e1}	the linear electrostatic spring coefficient of DETF
k_{e3}	the cubic electrostatic spring coefficient of DETF
k_{ed}	the elastic coefficient of the squeeze film air damping of DETF
k_{m3}	the cubic mechanical spring coefficient of DETF
k_{sus}	the transverse stiffness of suspensions
k_{vc}	the axial stiffness of connection beam of micro-lever
k_{vp}	the axial stiffness of pivot beam of micro-lever
k_{vt}	the axial stiffness of DETF tine beam
$k_{x,Single}$	the flexure stiffness of single beam suspensions on x axis
$k_{y,Single}$	the flexure stiffness of single beam suspensions on y axis
$k_{z,Single}$	the flexure stiffness of single beam suspensions on z axis
L_B	the length of flexure beams of micro-lever
L_{DETF}	the tine length of DETF sensing element
L_{con}	the length of connection beam of micro-lever
L_{Elc}	the electrode length of parallel-plate transducers
$L_{Elc,\Delta T}$	the electrode length of thermal expanded parallel-plate transducers
L_{eq}	the equivalent inductance of the linear electrical circuit model of DETF
L'_{eq}	the equivalent inductance in the electrical circuit model of DETF after included mechanical and electrostatic nonlinear effects
$L_{p \cdot k\_x}$	the length of horizontal segment of parallel kinematic beams suspensions
$L_{p \cdot k\_y}$	the length of vertical segment of parallel kinematic beams suspensions
L_{Single}	the length of single beam suspensions

L_T	the length of DETF tine beam
$L_{T,\Delta T}$	the length of thermal expanded DETF tine beam
$L_{h_{Snake}}$	the length of horizontal segment of snake beam suspensions
$L_{v_{Snake}}$	the length of vertical segment of snake beam suspensions
M_{eff}	the effective mass of DETF tine beam
M_{Proof}	the proof mass of MEMS RXL
$Mod\sigma_y(\tau)$	the modified Allan Variance
m_{Ele}	the mass of attached electrodes on DETF tine beam
N_{Box}	the smoothing number of BOXCAR filter
NB	the white noise power in the bandwidth B
$n(t)$	the near-DC noises in an amplifier
P	the electrostatic force between the electrodes of electromechanical transducers
P_0	the power of carrier signal
P_a	the ambient pressure
P_c	the power required to maintain the oscillation amplitude of DETF
Q	the quality factor of DETF sensing element
	the total charge stored across a capacitor (Section 2.3.1)
q	the electron charge
$|q_1|$	the maximum vibration amplitude of DETF vibrating at the fundamental resonant frequency
$q_i(t)$	the modal coordinate of the i-th mode of C-C beam
R_{eq}	the equivalent resistance in the linear electrical circuit model of DETF
R'_{eq}	the equivalent resistance in the electrical circuit model of DETF after included mechanical and electrostatic nonlinear effects
R_s	the series resistance in the electrical circuit model of DETF
R_{TIA}	the gain resistance of trans-impedance amplifier
S_{21}	the measured forward transmission coefficient of DETF sensing element
$S_{\alpha\_AC}(f)$	the one-side PSD of the voltage amplitude noise of AC driving signal
$S_{\Delta v}(f)$	the one-side PSD of frequency noise

$S_\varphi(f)$	the one-side PSD of phase noise
$S_{\varphi\_amp}(f)$	the one-side PSD of the amplifier-induced phase noise
$S_{\varphi\_thermal}(f)$	the one-side PSD of mechanical-thermal noise-induced phase noise
$S_{\varphi\_v-f}(f)$	the one-side PSD of phase noise induced by nonlinear voltage-frequency conversion effect
$S_\Psi(f)$	the one-side PSD of phase noise without considering Leeson Effect (Section 3.3)
S_{Axl}	the scale factor of MEMS RXL
S'_{Axl}	the scale factor of the MEMS RXL with two differential DETFs (Section 5)
$S_{DC}(f)$	the one-side PSD of the voltage noise of DC bias voltage
S_{Res}	Sensitivity/Scale factor of DETF sensing element
$S_y(f)$	the one-side PSD of fractional frequency noise
T	the temperature
$T.E.D_{Lvr,in}$	the thermal expansion-induced displacement at the input end of micro-lever
$T.E.D_{Lvr,out}$	the thermal expansion-induced displacement at the output end of micro-lever
t	Time
	the thickness of device layer (Section 2.2)
t_B	the thickness of the flexure beams of micro-lever
t_{Elc}	the electrode thickness of parallel-plate transducers
$t_{Elc,\Delta T}$	the electrode thickness of thermal expanded parallel-plate transducers
t_{Single}	the thickness of single beam suspensions
t_T	the thickness of the DETF tine beam
$t_{T,\Delta T}$	the thickness of thermal expanded DETF tine beam
U	the total energy stored in a system
u_x	the displacement on x axis
u_y	the displacement on y axis
u_z	the displacement on z axis
V_0	the voltage cross a capacitor (Section 2.3)
V_P	the DC polarization voltage applied on the electrodes
$V_{Thermal\_noise}$	the electronic thermal noise voltage
$v(t)$	a time-varied voltage signal
$\bar{v}(t)$	a time-varied voltage signal with phase noise
$v_0(t)$	the output voltage signal of an amplifier (Section 3.3.2)

v_d	the AC drive voltage applied on the electrodes
w_B	the width of flexure beams of micro-lever
w_{DETF}	the tine width of DETF sensing element
$w_{p.k\_x}$	the width of horizontal segment of parallel kinematic beams suspensions
$w_{p.k\_y}$	the width of vertical segment of parallel kinematic beams suspensions
w_{Single}	the width of single beam suspensions
w_T	the width of the DETF tine beam
$W_{T.\Delta T}$	the width of thermal expanded DETF tine beam
w_h_{Snake}	the width of horizontal segment of snake beam suspensions
w_v_{Snake}	the width of vertical segment of snake beam suspensions
$|x_0|$	the maximum displacement of DETF beams vibrating at its resonant frequency
x_{Lvr_in}	the displacement at the input end of micro-lever
$\bar{y}_k$	the k-th contiguous measurement sample of oscillator output frequency
Z_1	the equivalent input impendence
Z_2	the equivalent output impendence
Z_r	the impendence of a resonator

1

Mechanical Design of Micromechanical Silicon Oscillating Accelerometer

Anping Qiu

No. 200 Xiaolingwei, Xuanwu District, Nanjing, China
E-mail: apqiu@njust.edu.cn

This chapter describes the mechanical design of the sensing element in MEMS silicon oscillating accelerometers (SOA). The fundamental operation principle of the SOAs is first presented, and the analytic solutions for critical parameters, natural frequency and force sensitivity of the DETF resonator are obtained. Then, focusing on the micro lever structure in the SOA, the micro lever mechanism is analysed, and the amplification factor of the micro lever and the system amplification factor are derived with the consideration of the effect from the support beams of the sensing element. Based on these analyses, the fundamental performance parameters for the SOA, including scale factor, nonlinearity and bias, are presented. Finally, the two major error sources that impact the SOA performance, e.g. thermal sensitivity and stiffness nonlinearity, are analysed, and the sensitivity of the bias to temperature and nonlinearity stiffness of the DETF resonator are discussed.

1.1 Introduction

Micromechanical silicon oscillating accelerometer (SOA) belongs to a generic category of accelerometers known as vibration beam accelerometer (VBA), whose direct output is a frequency related to the input acceleration. In most SOAs, the sense element consists of a proof mass and two double-ended tuning fork (DETF) resonators. A schematic representation is shown in Figure 1.1. The DETF resonators are joined via micro levers to the proof mass. Each of the DETF resonators is formed out of two flexure beams,

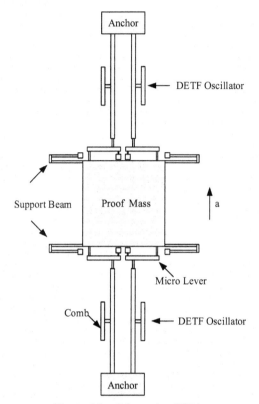

Figure 1.1 Schematic of SOA.

which are driven at resonance using an electrostatic comb-drive system. When accelerated along the input axis, the proof mass exerts a force on the forks, causing their natural frequencies to shift in opposing directions. Taking the difference of these two frequencies gives a measurement of the input acceleration.

The proof mass is attached to the substrate via anchor beams. The beams are flexible in the direction of the input axis, but are rigid along the other axes. In this way, only motion along the input axis is allowed.

These devices have been considered attractive for a number of reasons, including simpler dynamics and control, improved stability, large dynamic range, high resolution, and a quasi-digital frequency modulation (FM) output [1]. In contrast to those the currently available commercial silicon accelerometers that translate input acceleration into quasi-DC quantities, such as capacitance or resistance, the SOA translates the input acceleration into a change of mechanical vibration and is therefore less susceptible to the power supply and temperature fluctuations [2].

1.2 Mechanical Structure Design

1.2.1 Theory of Operation

As shown in Figure 1.2, the SOA can be divided into five major parts: a proof mass, two micro levers and two DETF resonators. The proof mass is suspended between two micro levers. Two DETF resonators sense the acceleration through the changes in their nature frequencies induced by the acceleration of the proof mass via force amplification mechanism of the micro lever.

The two DETF resonators are kept vibrating at its natural frequency by electrostatic forces. The configuration of the two DETF resonators causes the accelerating proof mass to load one DETF resonator's beams in tension while at the same time loading the other in compression. The tension loading increases the natural frequency of one DETF resonator, and the compression loading causes the natural frequency of the other DETF resonator to decrease. The difference in the resonance frequencies of the two DETF resonators is a measure of the acceleration along the input axis of the SOA [3].

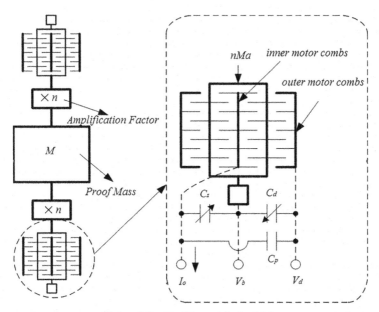

Figure 1.2   Simple model of a SOA.

The differential structure design improves the sensitivity of SOA and also effectively suppresses common-mode errors of the two DETF resonators. Employing this common-mode rejection technique greatly reduces the frequency shift caused by common-mode errors, such as temperature, off-axis accelerations and nonlinearity of oscillation [3].

By adopting a micro lever mechanism, the size of the proof mass is greatly decreased while holding high sensitivity. Mechanical transformation in a micro lever mechanism is achieved by elastic deformation of its component flexure beams. The design of the micro lever mechanisms is different from that of the conventional levers in the macro-world. A pivot structure in the macro-world can be formed by a pin-joint or bearing that permits free rotation and rigid support. With fabrication technology constraints, it is very difficult to achieve free rotation and rigid support in a micro lever mechanism. The micro lever mechanism is mainly formed by co-planar flexures. The most commonly used pivot in MEMS is a flexure beam with one side anchored to the substrate as a pseudo-pivot [4]. The geometry of the beam is designed to have a relatively small rotational spring constant allowing easy rotation. A single-stage micro lever mechanism is used to magnify the inertial force, which consists of four major parts: lever arm, pivot, the input system and output system [4].

1.2.2 Modelling of DETF Resonator for Closed-form Analysis

As illustrated in Figure 1.3, the resonator contains two beams and a comb structure located in the middle of each DETF resonator. The two DETF resonators vibrate 180° out of phase. The interaction of one DETF resonator with the other is determined through the elastic boundary conditions. Since the maximum vibration displacement of the DETF resonator is usually less than

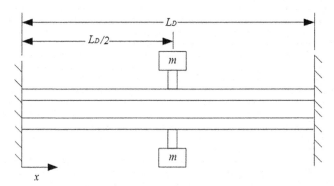

Figure 1.3 Model of the DETF resonator.

1 μm, the elastic boundary conditions of the DETF resonator are assumed to be fixed-fixed boundary conditions. Transverse vibration of the DETF resonator is governed by the Bernoulli–Euler equation [5].

The equation of the DETF resonator with fixed-fixed boundary conditions is as follows:

$$\rho A_D \frac{\partial^2 w}{\partial t^2} - F \frac{\partial^2 w}{\partial x^2} + EI \frac{\partial^4 w}{\partial x^4} = P_v \tag{1.1}$$

$$w(0,t) = \frac{\partial w(0,t)}{\partial x} = 0$$

$$w(L_D,t) = \frac{\partial w(L_D,t)}{\partial x} = 0$$

where $w(x,t)$ is the transverse displacement, A_D the cross-sectional area of the DETF resonator, F the axial load, I the moment of inertia of cross section of the DETF resonator. P_v is the electrostatic force exerted on comb drive actuator and can be equivalent to a concentrated load as follows [6]:

$$P_v = -m\ddot{w}\left(\frac{L_D}{2}\right)\delta\left(\frac{L_D}{2}\right) + P\delta\left(\frac{L_D}{2}\right) \tag{1.2}$$

The solution of Equation (1.1) begins using the technique of separation of variables. Let:

$$w(x,t) = \sum_i \phi_i(x)q_i(t) \tag{1.3}$$

where $\phi_i(x)$ represents the i th mode shape, a function of x alone, and $q_i(t)$ indicts how the amplitude of the shape varies with time t.

Inserting Equations (1.2) and (1.3) into Equation (1.1) gives:

$$\sum_i \left(EI \frac{\partial^4 \phi_i}{\partial x^4} q_i - F \frac{\partial^2 \phi_i}{\partial x^2} q_i + \rho A \phi_i \ddot{q}_i \right.$$

$$\left. + m\phi_i \left(\frac{L_D}{2}\right) \delta \left(\frac{L_D}{2}\right) \ddot{q}_i \right) = P\delta \left(\frac{L_D}{2}\right) \tag{1.4}$$

Considering modal orthogonality, the independent modal equations can be obtained:

$$\left(\int_0^L \rho A_D \phi_i^2 dx + m\phi_i^2 \left(\frac{L_D}{2}\right) \right) \ddot{q}_i$$

$$+ \left(\int_0^L EI \left(\frac{\partial^2 \phi_i}{\partial x^2}\right)^2 dx + \int_0^L F \left(\frac{\partial \phi_i}{\partial x}\right)^2 dx \right) q_i = P\phi_i \left(\frac{L_D}{2}\right) \tag{1.5}$$

Introducing $\varepsilon = x/L_D$, Equation (1.5) can be rewritten as:

$$M_{eff}\ddot{q}_i + K_{eff}q_i = P_{eff} \tag{1.6}$$

where

$$M_{eff} = \rho A_D L_D \int_0^1 \phi_i^2 d\varepsilon \; + \; m\left(\phi_i\left(\frac{1}{2}\right)\right)^2$$

$$K_{eff} = \frac{EI}{L_D^3}\int_0^1 \left(\frac{d^2\phi_i}{d\varepsilon^2}\right)^2 d\varepsilon + \frac{F}{L_D}\int_0^1 \left(\frac{\partial\phi_i}{\partial\varepsilon}\right)^2 d\varepsilon \tag{1.7}$$

$$P_{eff} = P\phi_i\left(\frac{1}{2}\right)$$

where the natural frequency of the DETF resonator can be expressed as:

$$\omega_0 = \sqrt{\frac{K_{eff}}{M_{eff}}} \tag{1.8}$$

Next, to solve for the mode shapes, the partial differential equation of free vibration of the DETF resonator is as follows:

$$\rho A_D \frac{\partial^2 w}{\partial t^2} + EI \frac{\partial^4 w}{\partial x^4} = 0 \tag{1.9}$$

Inserting Equation (1.3) into (1.9) yields:

$$\sum_i a^2 \frac{\partial^4 \phi_i}{\partial x^4}\cdot\frac{1}{\phi_i} = \sum_i -\frac{\partial^2 q_i}{\partial t^2}\cdot\frac{1}{q_i} \tag{1.10}$$

where $a^2 = EI/\rho A_D$. Since the right hand side of the equation depends only on x and the left hand side only on t, both sides are equal to some constant value $\Sigma p^2 i$. Thus:

$$\frac{\partial^4 \phi_i}{\partial x^4} - \beta_i^4 \phi_i = 0 \tag{1.11}$$

where $\beta_i^4 = p_i^2/a^2$.

The general solution of Equation (1.11) is as follows:

$$\phi_i(x) = A\sin\beta_i x + B\cos\beta_i x + Csh\beta_i x + Dch\beta_i x \tag{1.12}$$

With fixed-fixed boundary conditions,

$$\phi_i(0) = \phi_i'(0) = 0$$
$$\phi_i(L_D) = \phi_i'(L_D) = 0$$

Hence, the shape functions can be obtained:

$$\phi_i(x) = D\left[(ch\beta_i x - \cos\beta_i x) - \frac{ch\beta_i L_D - \cos\beta_i L_D}{sh\beta_i L_D - \sin\beta_i L_D}(sh\beta_i x - \sin\beta_i x)\right]$$

(1.13)

Inserting $\varepsilon = x/L_D$ into Equation (1.13), the form is rewritten as follows:

$$\phi_i(\varepsilon) = D\left[\left(ch\left(i + \frac{1}{2}\right)\pi\varepsilon - \cos\left(i + \frac{1}{2}\right)\pi\varepsilon\right)\right.$$

$$-\frac{ch\left(i + \frac{1}{2}\right)\pi - \cos\left(i + \frac{1}{2}\right)\pi}{sh\left(i + \frac{1}{2}\right)\pi - \sin\left(i + \frac{1}{2}\right)\pi}$$

$$\left.\times\left(sh\left(i + \frac{1}{2}\right)\pi\varepsilon - \sin\left(i + \frac{1}{2}\right)\pi\varepsilon\right)\right] \quad (1.14)$$

Since the DETF resonator always operates in the first-order mode, letting $\phi_1(1/2) = 1$ and inserting into Equation (1.14), the values can be obtained as follows:

$$\int_0^1\left(\frac{d^2\phi_1}{d\varepsilon^2}\right)^2 = 198.6, \int_0^1\left(\frac{\partial\phi_1}{\partial\varepsilon}\right)^2 = 4.85, \int_0^1\phi_1^2 d\varepsilon = 0.397$$

Inserting the above values into Equation (1.7) yields:

$$M_{eff} = 0.397\rho w_D h L_D + m$$

$$K_{eff} = 16.55Eh\left(\frac{w_D}{L_D}\right)^3 + 4.85\frac{F}{L_D} \quad (1.15)$$

$$P_{eff} = P$$

where w_D is the width of the DETF resonator and h the depth of the DETF resonator. Hence, the natural frequency of the DETF resonator under an axial force can be expressed as follows:

$$f_n = \frac{1}{2\pi}\sqrt{\frac{K_{eff}}{M_{eff}}} = \frac{1}{2\pi}\sqrt{\frac{16.55Eh\left(\frac{w_D}{L_D}\right)^3 + 4.85\frac{F}{L_D}}{0.397\rho w_D h L_D + m}} \quad (1.16)$$

It is easy to see that the natural frequency of the DETF resonator without the axial force is as follows:

$$f_0 = \frac{1}{2\pi}\sqrt{\frac{K_0}{M_{eff}}} \quad (1.17)$$

where $K_0 = 16.55Eh(\frac{w_D}{L_D})^3$.

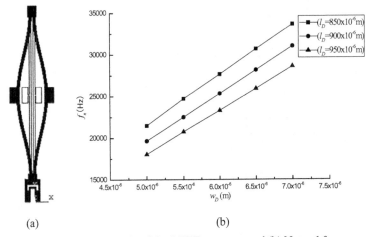

(a)   (b)

Figure 1.4   (a) First vibration mode of the DETF resonator and (b) Natural frequency versus width of the DETF resonator.

Figure 1.4(a) shows the vibration mode shape of the DETF resonator. Figure 1.4(b) displays the natural frequency of the DETF resonator as a function of vibrating beam width and length. This figure shows that the natural frequency increases with both increasing vibrating beam width and decreasing length. Since the SOA scale factor varies inversely with the natural frequency of the DETF resonator (seen in Section 1.2.4), and the narrowest vibrating beam width is limited to several microns by the current high-depth-width-ratio bulk fabrication processes, a general design value of the natural frequency of the DETF resonator is about 25 kHz.

1.2.3 Micro Lever Mechanism and Amplification Factor

In order to maximize the mechanical sensitivity available from a small proof mass, a lever system is used to magnify the force applied to the two DETF resonators [2]. Figure 1.5 shows the deformation schematic of the micro lever, and Figure 1.6 illustrates the applied forces and bending moments, as well as reactions, which act on the micro lever [7].

For this analysis, it is assumed that the lever arm is rigid and remains straight during loading. The total deflection can be represented by a vertical displacement δ and a rotation angle θ about the pivot. Applying the force and moment equilibrium condition to the lever arm leads to the following [6]:

$$F_{in} = k_{vvo}(l\theta + \delta) + k_{vvp}\delta \qquad (1.18)$$

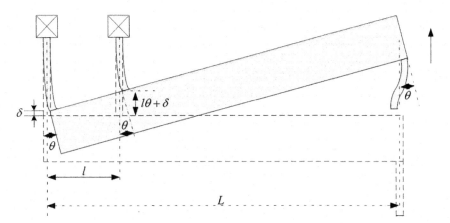

Figure 1.5 Deformation schematic of the micro lever.

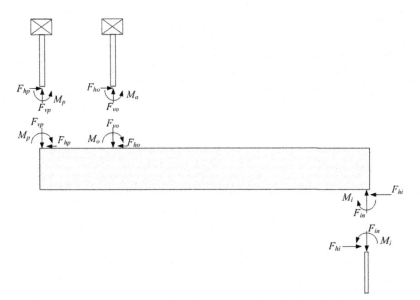

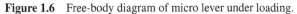

Figure 1.6 Free-body diagram of micro lever under loading.

$$F_{in}L = k_{vvo}(l\theta + \delta)l + M_o + M_p + M_i \tag{1.19}$$

$$F_{ho} + F_{hp} + F_{hi} = 0 \tag{1.20}$$

where k_{vvo} and k_{vvp} represent the axial stiffness at the output beam of the micro lever and the pivot beam, respectively; L is the length between the pivot beam and the input beam, l the length between the input beam and the output beam, M_o, M_p and M_i are the bending moment at the output beam, the pivot beam and the input beam, respectively; F_{ho}, F_{hp} and F_{hi} are the tangential forces at the output beam, the pivot beam and the input beam, respectively.

The rotation angle at the pivot beam θ is equal to the rotation angle of the rigid lever arm and separated into two components; θ_{mp} is caused by the bending moment M_p and θ_{hp} by the tangential force F_{hp}.

$$\theta = \theta_{mp} + \theta_{hp} \tag{1.21}$$

$$\theta_{mp} = \frac{M_p}{k_{\theta mp}} \tag{1.22}$$

$$\theta_{hp} = \frac{F_{hp}}{k_{\theta hp}} \tag{1.23}$$

where $k_{\theta mp}$ is the bending stiffness of the pivot when loaded with M_p and $k_{\theta hp}$ the bending stiffness of the pivot under F_{hp}.

The rotation angle at the input beam θ is equal to the rotation angle of the rigid lever arm and separated into two components, θ_{mo} caused by the bending moment M_o and θ_{ho} caused by the tangential force F_{ho}.

$$\theta = \theta_{mo} + \theta_{ho} \tag{1.24}$$

$$\theta_{mo} = \frac{M_o}{k_{\theta mo}} \tag{1.25}$$

$$\theta_{ho} = \frac{F_{ho}}{k_{\theta ho}} \tag{1.26}$$

where $k_{\theta mo}$ is the bending stiffness of the pivot when loaded with M_o, $k_{\theta ho}$ the bending stiffness of the pivot under F_{ho}.

The rotation angle at the input beam θ is equal to the rotation angle of the rigid lever arm and separated into two components, θ_{mi} caused by the bending moment M_i and θ_{hi} caused by the tangential force F_{hi}.

$$\theta = \theta_{mi} + \theta_{hi} \tag{1.27}$$

$$\theta_{mi} = -\frac{M_i}{k_{\theta mi}} \tag{1.28}$$

$$\theta_{hi} = \frac{F_{hi}}{k_{\theta hi}} \tag{1.29}$$

where $k_{\theta mi}$ is the bending stiffness of the input beam when loaded with M_i and $k_{\theta hi}$ the bending stiffness of the input beam under F_{hi}.

Replacing the total rotation angle θ in Equation (1.19) with the rotation angle caused by the bending moment, θ_{mp}, θ_{mo} and θ_{mi}, the following equation for the moment equilibrium with respect to the joint between the pivot beam and the lever arm are given by:

$$F_{in}L = k_{vvo}(l\theta + \delta)l + k_{\theta mo}\theta_{mo} + k_{\theta mp}\theta_{mp} - k_{\theta mi}\theta_{mi} \tag{1.30}$$

Let

$$\theta_{mp} = f_p\theta, \theta_{mo} = f_o\theta, \theta_{mi} = f_i\theta \tag{1.31}$$

Substituting Equation (1.31) into Equations (1.30) and (1.30) becomes:

$$F_{in}L = k_{vvo}(l\theta + \delta)l + k_{\theta mo}f_o\theta + k_{\theta mp}f_p\theta - k_{\theta mi}f_i\theta \tag{1.32}$$

Solving Equations (1.18) and (1.32), the equivalent axial stiffness K_{in} and the amplification factor n are given as follows:

$$
\begin{aligned}
K_{in} &= \frac{F_{in}}{(L\theta + \delta)} \\
&= \frac{(k_{vvo} + k_{vvp})(f_o k_{\theta mo} + f_p k_{\theta mp} - f_i k_{\theta mi}) + k_{vvo}k_{vvp}l^2}{(f_o k_{\theta mo} + f_p k_{\theta mp} - f_i k_{\theta mi}) + k_{vvo}(L-l)^2 + k_{vvp}L^2}
\end{aligned}
\tag{1.33}
$$

$$
n = \frac{k_{vvo}(l\theta + \delta)}{F_{in}} = \frac{\frac{1}{k_{vvp}}(f_o k_{\theta mo} + f_p k_{\theta mp} - f_i k_{\theta mi}) + IL}{\left(\frac{1}{k_{vvo}} + \frac{1}{k_{vvp}}\right)(f_o k_{\theta mo} + f_p k_{\theta mp} - f_i k_{\theta mi}) + l^2}
\tag{1.34}
$$

Next, to find the coefficients of f_p, f_o and f_i, first, using Equations (1.20) to (1.29), the following equation can be obtained:

$$k_{\theta hp}(1 - f_p) + k_{\theta ho}(1 - f_o) + k_{\theta hi}(1 - f_i) = 0 \tag{1.35}$$

Since the lever arm remains rigid during loading, assume that the horizontal displacements of the ends of the beams attached the lever arm are the same. The horizontal displacement can be divided into two parts, caused by tangential force and bending moment, respectively, as follows:

$$\gamma_{mo} + \gamma_{ho} = \gamma_{mp} + \gamma_{hp} = \gamma_{mi} + \gamma_{hi} \tag{1.36}$$

where γ_{mo}, γ_{mp} and γ_{mi} are the horizontal displacements of the output beam, pivot beam and input beam under the bending moments, respectively. γ_{ho},

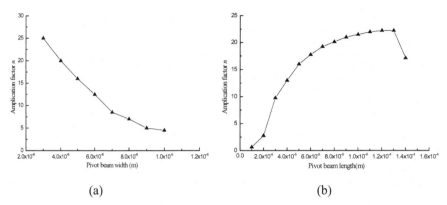

(a)   (b)

Figure 1.7   Amplification factor versus (a) pivot beam width and (b) pivot beam length.

γ_{hp} and γ_{hi} are the horizontal displacements of the output beam, pivot beam and input beam under the tangential forces, respectively. This gives:

$$\frac{k_{\theta mo}}{k_{hmo}}f_o\theta + \frac{k_{\theta ho}}{k_{hho}}(1-f_o)\theta = \frac{k_{\theta mp}}{k_{hmp}}f_p\theta + \frac{k_{\theta hp}}{k_{hhp}}(1-f_p)\theta$$

$$= \frac{k_{\theta mi}}{k_{hmi}}f_i\theta + \frac{k_{\theta hi}}{k_{hhi}}(1-f_i)\theta \qquad (1.37)$$

where k_{hmo}, k_{hmp} and k_{hmi} are the horizontal displacements of the output beam, pivot beam and input beam under the bending moments, respectively; k_{hho}, k_{hhp} and k_{hhi} represent the tangential stiffness of the output beam, pivot beam and input beam under the tangential forces, respectively.

Finally, using Equations (1.35) and (1.37), solve for f_p, f_o and f_i. Substituting the coefficients into Equations (1.32) and (1.34), the equivalent axial stiffness K_{in} and the amplification factor n are given.

Figure 1.7 shows the amplification factor n as a function of the size of the pivot beam. In Figure 1.7(a), the amplification factor n decreases with the increasing pivot beam width. Figure 1.7(b) shows that the amplification factor n increases with the increasing pivot beam length within a certain range. As the pivot beam length increases, the axial stiffness of the pivot beam k_{vvp} decreases. As seen from Equation (1.34), the axial deformation of the pivot beam results in a reduction of the amplification factor. Therefore, in order to obtain the maximum amplification factor, the design of the pivot beam is the key factor.

1.2.4 System Amplification Factor *n′*

When an acceleration along the input axis is applied to the SOA, the force from the proof mass is magnified by the micro lever and then transferred to the DETFs. The effective amplification factor n' is defined as the ratio of the axial force of the DETF beam to the input inertial force of the proof mass [6].

The mechanism of the structure of SOA can be reduced to the dynamic model of a single freedom mass-spring vibration system. Applying the force to the proof mass leads to the following equation:

$$4K_{in}x_a + 4K_m x_a = Ma \qquad (1.38)$$

where K_m is the bending stiffness of the input axis, M the proof mass of SOA and x_a the displacement of the proof mass caused by input acceleration. Hence, the system amplification factor is dependent upon the mechanical stiffness of the flexible beam in the accelerometer. Thus, the system amplification factor n' is:

$$n' = \frac{K_{in}}{K_{in} + K_m} n \qquad (1.39)$$

1.2.5 Scale Factor

When the proof mass of SOA experiences an input acceleration a, one DETF resonator is loaded in tension, while the other DETF resonator is loaded in compression. The amplitude of the axial force of the DETF resonators is:

$$F = \frac{n'Ma}{4} \qquad (1.40)$$

Substituting Equation (1.40) into Equation (1.16) yields:

$$f_n = f_0\sqrt{1 \pm \frac{\frac{4.85}{L_D}\left(\frac{n'Ma}{4}\right)}{K_0}} \qquad (1.41)$$

The difference in the natural frequencies of the DETF resonators is expressed as follows:

$$\Delta f = f_0\sqrt{1 + \frac{\frac{4.85}{L_D}\left(\frac{n'Ma}{4}\right)}{K_0}} - f_0\sqrt{1 - \frac{\frac{4.85}{L_D}\left(\frac{n'Ma}{4}\right)}{K_0}} \qquad (1.42)$$

Using the Taylor series, and neglecting the higher orders, Equation (1.42) can be rewritten as:

$$\Delta f \approx f_0\frac{4.85n'Ma}{4L_D K_0} + \frac{1}{8}f_0\left(\frac{4.85n'Ma}{4L_D K_0}\right)^3 \qquad (1.43)$$

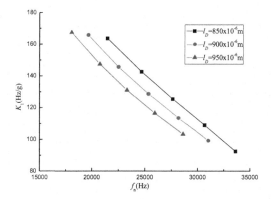

Figure 1.8   SOA scale factor versus natural frequency of the DETF resonator.

Thus, the scale factor of the SOA, K_1, is:

$$K_1 = \frac{4.85n'M}{16\pi^2 L_D M_{eff} f_0} \left(\frac{\text{Hz}}{\text{m/s}^2}\right) \qquad (1.44)$$

SOA scale factor is displayed in Figure 1.8 as a function of the natural frequency of the DETF resonator and the vibrating beam length. This figure shows that scale factor decreases with both increasing natural frequency of DETF resonator and the increasing vibrating beam length.

1.2.6 Bias

Bias of the SOA is defined as the output frequency difference in the two DETF resonators given no acceleration input, also referred to as offset in the literature. Bias should be reduced first to improve bias stability, particularly in the case where the bias stability is proportional to the initial bias. The bias value is determined by how closely the two DETF resonator frequencies match, mostly due to the mismatches of the dimensions and stresses induced by the die and through die packaging of the resonators.

Considering the effect of residual stress, Equation (1.17) can be written as follows:

$$f_0 = \frac{1}{2\pi} \sqrt{\frac{K_0 + 4.85\frac{\sigma_s w_D h}{L_D}}{0.397\rho w_D h L_D + m}} \qquad (1.45)$$

where σ_s is the applied thermal stress on the beams of the resonators. Considering the mismatch of the dimension and stress of the two resonators, the width mismatch Δw_D can be defined as $w_{D2} - w_{D1}$, and w_D can be defined as $\frac{w_{D1} + w_{D2}}{2}$. Similar definitions can be made for the stress σ_s. Using these definitions, the effect of change in dimension and stress on the natural frequency can be found by the analysis of Equation (1.45) [8]:

$$\frac{\Delta f_0}{f_0} = \frac{3\Delta w_D}{2w_D} - \frac{3\Delta L_D}{2L_D} + \frac{\Delta h}{h} + 2.43 \frac{w_D h}{K_0 L_D} \Delta \sigma_s \qquad (1.46)$$

For bulk micro fabrication, normally $\frac{\Delta L_D}{L_D}, \frac{\Delta h}{h} \ll \frac{\Delta w_D}{w_D}$, so Equation (1.46) can be simplified to:

$$\frac{\Delta f_0}{f_0} \approx \frac{3\Delta w_D}{2w_D} + 2.43 \frac{w_D h}{K_0 L_D} \Delta \sigma_s \qquad (1.47)$$

Thus:

$$B \approx \frac{f_0 \left(\frac{3\Delta w_D}{2w_D} + 2.43 \frac{w_D h}{K_0 L_D} \Delta \sigma_s \right)}{K_1} \qquad (1.48)$$

Note that the mismatches in the beam width and the applied residual stress tend to dominate the value of the bias. One way to minimize the width mismatch is to set the beam width much larger than the minimum line width of the fabrication process. However, this will decrease the sensitivity of the DETF resonator, and this trade-off must be considered. Meanwhile, The DETF resonators should be insulated from the residual stress in order to obtain a good common-mode error suppression for the SOA.

1.2.7 Thermal Sensitivity

Temperature error is a key error source that affects the performance of the SOA. The thermal sensitivity effect is divided into two counteracting components [3], thermal stresses due to differences in material expansions induced into SOA sensitivity element and the modulus of elasticity of the DETF resonator material. Both affect the frequencies of the resonators.

The effect of change in modulus and stress on the natural frequency can be found by the analysis of Equation (1.45)

$$\frac{1}{f_0} \frac{\partial f_0}{\partial T} \approx \frac{1}{2E} \frac{\partial E}{\partial T} + 2.43 \frac{w_D h}{K_0 L_D} \frac{\partial \sigma_s}{\partial T} \qquad (1.49)$$

The Young's modulus of silicon decreases with the increasing temperature. The decrease in the Young's modulus makes the natural frequencies of the resonators decrease since the resonator beam stiffness decreases. The temperature coefficient of the Young's modulus of silicon is about -25 to -75 ppm/°C [10]. This change corresponds to a downward shift in the natural frequencies of the resonators of -12.5 to -37.5 ppm/°C.

Assume that the sense element is anchored on a glass substrate. Since the glass substrate is much thicker than the silicon, the stiffness of the substrate is much greater than the silicon. The movement of the anchor points due to temperature changes would be driven solely by the glass substrate. While the thermal expansion coefficient of the glass substrate is greater than the silicon, the beams of the DETF resonators are in tension with the increasing temperature, and this causes an increase in the resonators' frequencies. Conversely, the beams of the DETF resonators are in compression with the increasing temperature, and this causes a decrease in the resonators' frequencies.

From Equation (1.50), it is easy to note that if two DETF resonators are exactly matched and subject to the exact same environment, the frequencies of the resonators will not shift with respect to each other. However, the accuracy of the lithography-based process is on the order of 10^{-2} to 10^{-3} (the ratio of the average defect to the smallest feature size), while the accuracy of conventional manufacturing utilizing precision machining is two to three orders of magnitude higher, on the order of 10^{-5} [11].

According to Equation (1.48), the sensitivity of the bias to temperature can be modelled as:

$$B_T \approx f_0 \begin{bmatrix} \left(\frac{3}{4Ew_D}\frac{\partial E}{\partial T} + 3.65\frac{h}{K_0L_D}\frac{\partial \sigma_s}{\partial T} \right)\Delta w_D \\[2mm] +2.43\frac{w_D h}{K_0L_D}\left(\frac{1}{2E}\frac{\partial E}{\partial T} + 2.43\frac{w_D h}{K_0L_D}\frac{\partial \sigma_s}{\partial T} \right)\Delta \sigma_s \\[2mm] +2.43\frac{w_D h}{K_0L_D}\frac{\partial \Delta \sigma_s}{\partial T} \end{bmatrix} \qquad (1.50)$$

It can be seen from Equation (1.50) that due to the mismatches of the dimension and the applied stress, the common-mode error caused by temperature cannot be well suppressed. In order to minimize B_T error, it is necessary to improve the fabrication tolerance of the width of the beams of the DETF resonators and insulate against the thermal stresses delivered to the resonators.

1.2.8 Stiffness Nonlinearity

Stiffness nonlinearity of the DETF resonator caused by the nonlinear restoring force is that the springs stiffen with the displacement amplitude, and thus the resonant frequency is increased. Then the undamped motion equation for the DETF resonator is given as:

$$M_{eff}\ddot{q} + c\dot{q} + K_{eff}q + K_3q^3 = P \tag{1.51}$$

Considering small-amplitude vibration, the resonant frequency can be expressed as:

$$\omega^2 = \omega_n^2 \left(1 + \frac{3}{4}\frac{K_3}{K_{eff}}|q|^2\right) \tag{1.52}$$

where K_3 is nonlinearity stiffness and $|q|$ is the vibration amplitude of the DETF resonator.

Then the frequency change with displacement amplitude is as follows:

$$\frac{\Delta\omega}{\omega_n} = \frac{3}{8}\frac{K_3}{K_{eff}}|q|^2 \tag{1.53}$$

Thus, improving the amplitude control of the DETF resonators helps decrease bias uncertainty, and lowering the vibration amplitude can significantly reduce this error.

Next, solve for K_3. Figure 1.9(a) shows one-fourth of the mechanical model of the SOA, considering the effects of the micro lever and the support beams. The beam of the DETF resonator is simplified axially and constrained at both ends, with elastic constraints represented by stiffness K_a at one end (see Figure 1.9 (b)).

The model here adopted for the dynamic response of the beam is restricted by several hypotheses: (i) the beam is modelled by the Euler–Bernoulli theory, (ii) variation of the cross section during vibration is neglected, and (iii) stretching of the beam is small but finite. With the above hypotheses, the axial strain along the beam axis is [6, 9]:

$$\frac{du}{dx} = \frac{N - F_0}{EA_D} - \frac{1}{2}\left(\frac{dw}{dx}\right)^2 \tag{1.54}$$

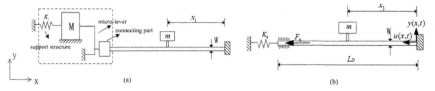

Figure 1.9 (a) One-fourth of the mechanical model of SOA and (b) the beam model.

Integrating in space Equation (1.54) with the boundary conditions:

$$u\left(0,t\right) = 0, u\left(L_D,t\right) = -\frac{(N - F_0)}{K_a}$$

The axial force can be given by:

$$N = F_0 + \frac{1}{2}\frac{K_a E A_D}{K_a L + E A_D}\int_0^L \left(\frac{dw}{dx}\right)^2 dx \tag{1.55}$$

Note that the axial force has two contributions: the first one is due to a pre-stress F_0/A_D acting on the beam of the DETF resonator, constant in time and independent from the beam transversal displacement w, while the second one is generated by the elongation of the beam induced by its finite deflection. This second contribution is present only in axially constrained beams.

Substituting the axial force into Equation (1.1) and referring to derivation of the formula of the natural frequency under axial force yields:

$$K_3 = \frac{1}{2}\frac{K_a E A_D}{K_a L_D + E A_D}\frac{1}{L_D^2}\left(\int_0^1 \left(\frac{\partial\phi_1}{\partial\varepsilon}\right)^2 d\varepsilon\right)^2$$

$$= 11.76\,\frac{K_a E A_D}{K_a L_D + E A_D}\frac{1}{L_D^2} \tag{1.56}$$

in which

$$K_a = \frac{F_0}{x_0}$$

$$F_0 = n K_{in}\tau$$

$$x_o = \frac{(\delta + l\theta)\tau}{\delta + L\theta}$$

such that

$$K_a = \frac{F_o}{x_o} = \frac{n K_{in}\left(L + \dfrac{\delta}{\theta}\right)}{l + \dfrac{\delta}{\theta}} \tag{1.57}$$

Since K_a is the elastic constraint stiffness of the beam of the DETF resonator, the DETF resonator is not taken into account when calculating n, K_{in}, δ and θ. Therefore, k_{vvo}, $k_{\theta mo}$ and $k_{\theta ho}$ in Equations (1.18) to (1.37) are replaced by the axial stiffness k'_{vvo}, the bending stiffness $k'_{\theta mo}$ under bending moment and the bending stiffness $k'_{\theta ho}$ under tangential force of the output beam, respectively. Using Equations (1.18) to (1.37), solve for the values of n, K_{in}, δ and θ and obtain the value of K_a. Substituting the value of K_a into Equation (1.56), the value of K_3 is obtained.

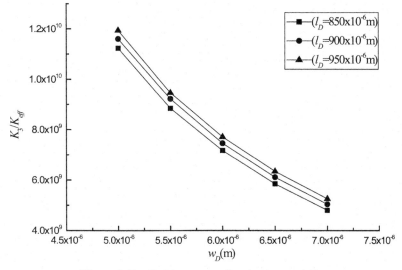

Figure 1.10 K_3/K_{eff} versus vibrating beam width.

Figure 1.10 shows the ratio of K_3 to K_{eff} as a function of the vibrating beam size of DETF resonator. As observed from the curves, the ratio of K_3 to K_{eff} is more sensitive to the vibrating beam width of the DETF resonator than the vibrating beam length. The ratio, K_3/K_{eff}, ranges from 10^9 to 10^{10} near a nominal natural frequency of 25 kHz. Substitution of these ratio values along with the natural frequency of 25 kHz and vibration amplitude of 0.1 μm into (1.53) yields a frequency change between 0.1 and 1 Hz. Assuming a common mode rejection factor of ten and an SOA scale factor of 100 Hz/g, the calculated bias stability is 0.1 to 1 mg due to the vibrating beam stiffness nonlinearity. A smaller oscillation amplitude would produce a smaller bias stability, but also a smaller output voltage signal from the capacitance sensing preamplifier [3].

1.3 Fabrication and Testing

1.3.1 SOA Fabrication

The SOA dies are fabricated using a bulk silicon-on-insulator (SOI) process and a wafer-level vacuum packaging compatible with the SOI process. A cross−section of a typical SOI wafer is shown in Figure 1.11. The die is realized with three wafers: the substrate, the SOI device layer and the cover.

The substrate wafer is shown in Figure 1.12(a). Oxidation of the substrate wafer is required to provide electrical signal isolation, mechanical anchors

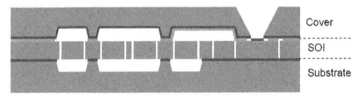

Figure 1.11 Cross section of the SOA dies.

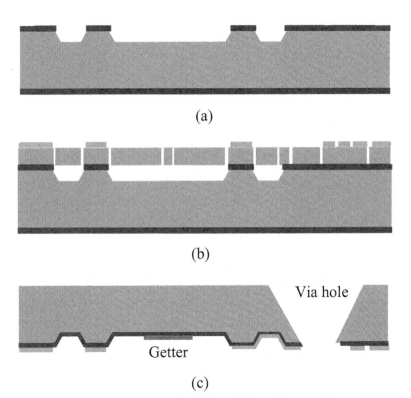

(a)

(b)

(c)

Figure 1.12 Cross sections of the wafers used to fabricate the sense element in the SOA.

and bond pads. After patterning the oxide layer, a photoresist is coated on the substrate wafer and patterned using the cavity mask. The cavities in the substrate wafer are dry-etched.

Then the SOI wafer and the substrate wafer are bonded by a silicon direct bond (SDB). The thickness of the SOI device layer is chosen such that it will match the desired proof mass thickness, typically 60 to 80 μm. Then, the deposit and etching of a metallic layer in the device layer forms the

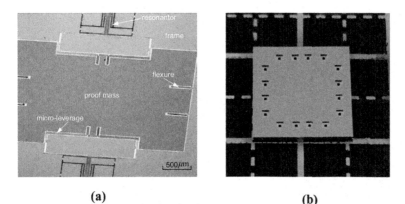

(a) **(b)**

Figure 1.13 (a) SEM image of the SOA, and (b) the wafer-level vacuum-packaged die.

interconnections for the electrical function of the device and bond pads for wire bonding. The most difficult processing is shown in Figure 1.12(b). In this step, a deep inductively-coupled plasma (ICP) anisotropic dry etch is performed, which defines the basic mechanical structure of the SOA. The high aspect ratio of ICP etch is up to 1:30. It is common to deposit metal in the cavity in the substrate wafer to minimize device layer undercut.

The cover wafer is shown in Figure 1.12(c). After cleaning, the cavities in the cover wafer are wet-etched. Then the cover is thermally oxidized with a thick oxide layer to provide the electrical isolation required between the cover and the device layer. Depositing and etching of a metallic layer in the cover wafer is then carried out to form the metal wiring and the sealing ring by evaporation or electroplating for Au/Si eutectic bonding. The via holes in the cover wafer are wet-etched for pad bonding.

Finally, the cover and the active SOI wafer are assembled by an Au/Si eutectic bonding (cf. Figure 1.11), forming a hermetic cavity which maintains the vacuum needed for high Q factor of the DETF resonator. In order to maintain the vacuum level over the long term, a getter is adhered to the inner surface of the cover wafer, and once activated, the getter progressively absorbs and traps gaseous species. The scanning electron micrograph (SEM) image of the SOA is shown in Figure 1.13(a) and the wafer-level vacuum-packaged SOA die in Figure 1.13 (b).

1.3.2 Testing

The fabricated SOA dies are attached to a Ceramic Leadless Chip Carrier package, as shown in Figure 1.14. The SOA is excited into self-resonance

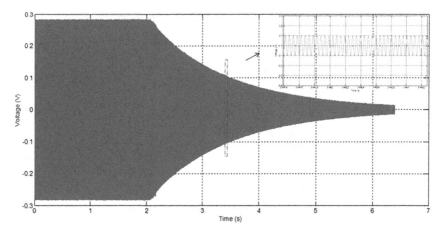

Figure 1.14 Amplitude decay of DETF resonator during ring-down tests.

Table 1.1 Drive frequencies and scale factors of each DETF resonator of the six devices

	f_1(kHz)*	SF_1(Hz/g)*	f_2 (kHz)*	SF_2(Hz/g)*	f (kHz)**	SF(Hz/g)**
1	19.898	76.5	20.291	74.0		
2	20.089	85.5	19.495	87.1		
3	19.808	78.4	20.651	71.5	19.560	90.7
4	19.46	86.5	19.283	87.2		
5	19.001	81.7	18.759	79.5		
6	20.145	84.1	19.515	79.3		

*Measures of the fabricated SOA
**Design values

along the vibration mode of the DETF resonator. The resonant frequency is acquired by a FPGA-based frequency measurement circuit. The frequency difference between the two DETF resonator channels is extracted as a measure of the acceleration. The drive frequencies and scale factors of each DETF resonator of the six fabricated devices are shown in Table 1.1.

Taking the ratio of the amplitude at resonance to the static deflection, the quality factor of a lightly damped DETF resonator reduces to

$$Q = \frac{1}{2\varsigma}$$

where ς is the damping ratio for the DETF resonator. The Q factor of the DETF resonator is experimentally characterized using ring-down tests, where the device is given an initial impulse and allowed to decay over time. Exponential fits of the time domain amplitude decay data reveal the Q factor of about 86,000, as shown in Figure 1.14. The bias stability of the SOA depends directly on the frequency stability of each DETF resonator and thus requires a high quality factor.

1.4 Conclusion

This chapter has presented the fundamental mechanical design of the MEMS sensing element in the SOA. The analytic solutions of critical quantities of the DETF resonator and micro lever mechanism have been derived. The fundamental performance parameters for the SOA have also been presented. The effects of temperature and stiffness nonlinearity on the SOA performance are analysed. Typical test results of the fabricated SOA sensing element dies show that the MEMS DETF resonator has a drive frequency $\sim$20 kHz and a scale factor of $\sim$80 Hz/g, with a Q factor up to 86,000.

References

[1] Trey A. Roessig, Roger T. Howe, Albert P. Pisano, and James H. Simth, 'Surface-micro-machined resonant accelerometer', International Conference on Solid-State Sensors and Actuators, pp. 859–862, Chicago, 1997.

[2] Lin He, Yong-ping Xu, and Anping Qiu, 'Folded silicon resonant accelerometer with temperature compensation', The 3rd IEEE Conference on Sensors, Vienna, Austria, October 2004.

[3] Kevin A. Gibbons, 'A micromechanical silicon oscillating accelerometer', Master thesis, the Massachusetts Institute of Technology, 1997.

[4] Xiao-Ping Susan Su, 'Compliant leverage mechanism design for MEMS applications', Ph.D. dissertation, Univ. California, Berkeley, CA, 2001.

[5] Trey Allen William Roessig, 'Integrate MEMS tuning fork oscillators for sensor applications', Ph.D. dissertation, Univ. California, Berkeley, CA, 1998.

[6] Ran Shi, 'Research on Key Technologies of Silicon Resonant Accelerometer', Ph.D. dissertation, Nanjing University of Science and Technology, Nanjing, China, 2013.

[7] Jing Zhang et. al., 'Micro-electro-mechanical resonant accelerometer designed with a high sensitivity', Journal of Sensors, Vol. 15, 2015.

[8] Jinhu Dong, 'Temperature characteristic of silicon resonant accelerometer', Master Thesis, Nanjing University of Science and Technology, Nanjing, China, 2012.

[9] Jing Zhang, Shao-dong Jiang, Qin Shi, Anping Qiu, 'Modelling of nonlinear stiffness of micro-resonator in silicon resonant accelerometer', Key Engineering Materials, Vol. 562–565, 2013.

[10] Matthew A. Hopcroft, William D. Nix, and Thomas W. Kenny, 'What is the Young's Modulus of Silicon?', Journal of Microelectromechanical Systems, 19(2): pp. 229–238, 2010.

[11] Andrei M. Shkel, 'Precision Navigation and Timing Enabled by Micro technology: Are We There yet?', Proc. of SPIE, Vol. 8031, 2011.

2

Front-end Amplifiers for MEMS Silicon Oscillating Accelerometers

Yang Zhao[1] and Yong Ping Xu[2]

[1]Nanjing University of Science & Technology, Nanjing, China
[2]Department of Electrical & Computer Engineering, National University of Singapore, Singapore
E-mail: zhaoyang0216@njust.edu.cn; yongping@ieee.org

In this chapter, the designs of the front-end amplifiers for MEMS silicon oscillating accelerometers (SOA) are described. This chapter first introduces the capacitive sensing principle for MEMS resonators, followed by a review of the front-end topologies adopted in MEMS oscillators since they are similar to the readout circuit in MEMS SOAs. The review covers single-stage and two-stage resistive transimpedance amplifier (TIA), T-network TIA, charge-sensing amplifier, capacitive feedback TIA, and focuses on their design trade-offs and techniques to relax these trade-offs. Three front-end amplifiers employed in the MEMS SOAs, having different topologies, are then described in detail. This chapter concludes with a performance comparison of some typical front-end amplifiers employed in MEMS oscillators and SOAs.

2.1 Capacitive Sensing in MEMS Sensors

Capacitive sensing is widely adopted in MEMS sensors since it doesn't require integration of heterogeneous materials and is compatible with almost all MEMS fabrication processes [1, 2]. A front-end amplifier is required to serve as an interface between an MEMS sensor and its readout circuit. It amplifies the sensor output, usually a very weak signal, for further signal conditioning. In the case of an MEMS resonator, the front-end amplifier detects a motion current signal induced by a capacitance change of the sense electrode, as shown in Figure 2.1 (a differential sensing topology is usually used and only a single-ended schematic is shown here for simplicity).

The difference between the above two detection methods is how the resonant beam is polarized. In Figure 2.1(a), a constant DC polarization voltage V_p is applied to the resonant beam. For the same capacitance variation, high V_p gives large motion current I_s, and hence a better signal-to-noise ratio (SNR). The static capacitance of the sense comb electrode C_s is

$$C_{s0} = 2n_s \frac{\varepsilon L_0 h}{g_0} \tag{2.1}$$

where n_s, L_0, h, and g_0 represent the number of fingers, overlap length, thickness, and the gap of the sense comb electrode, respectively. If the resonant beam displacement under the drive signal, $V_d(t)$, is x(t), as shown in Figure 2.1, C_s can be expressed as:

$$C_s(t) = 2n_s \frac{\varepsilon \left[L_0 + x(t) \right] h}{g_0} = C_{s0} + 2n_s \frac{\varepsilon h}{g_0} x(t) \tag{2.2}$$

The displacement of the resonator beam causes $C_s(t)$ to change and induces a motion current under the polarization voltage V_p. The resultant motion current is given by:

$$I_s(t) = \frac{dQ(t)}{dt} = V_p \frac{dC_s(t)}{dt} = V_p \frac{C_{s0}}{L_0} \cdot \frac{dx}{dt} \tag{2.3}$$

Clearly, the motion current is proportional to the velocity of the resonator beam movement. The front-end amplifier in Figure 2.1(a) senses and converts the motion current to a voltage, i.e., $V_o(t)$. Depending on the choice of R and C, the output voltage, $V_o(t)$, can be proportional to the velocity of the resonant beam, when $\omega_0 << 1/R_f C_f$, i.e.,

$$V_o(t) = -I_s(t) R_f \tag{2.4}$$

When $\omega_0 >> 1/R_f C_f$, the motion current, $I_s(t)$, is integrated onto C_f and the output voltage, $V_o(t)$, is given by

$$V_o(t) = \frac{-1}{C_f} \int I_s(t) \, dt = V_p \frac{C_{s0}}{L_0} x(t) \tag{2.5}$$

Now the output voltage, $V_o(t)$, is proportional to the displacement of the resonant beam, this is also referred to as the "charge-sensing amplifier"

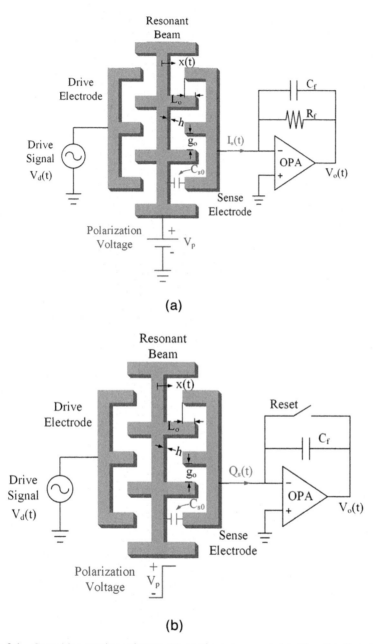

(a)

(b)

Figure 2.1 Capacitive sensing schemes (a) continuous current detection (b) discrete-time charge detection.

(CSA), as it effectively transfers the charge induced by the sense capacitance variation to C_f. In terms of the output phase, there is a 90^0 phase difference between these two amplifier topologies.

In Figure 2.1(b), a step polarization voltage is applied to the resonant beam, under which the capacitance change, $\Delta C_s(t)$, is converted to a charge signal, $\Delta Q_s(t)$, and is then detected by the front-end amplifier, i.e.,

$$Q_s(t) = Q_{s0} + \Delta Q_s(t) = -C_f V_o(t) \tag{2.6}$$

and

$$V_o(t) = -\frac{Q_{s0}(t)}{C_f} - \frac{\Delta Q_s(t)}{C_f} = -V_p\left[\frac{C_{s0}(t)}{C_f} + \frac{\Delta C_s(t)}{C_f}\right] \tag{2.7}$$

where assuming that the charge on C_f is reset before each V_p step.

The first static capacitance term in (2.7) can be removed by employing a differential topology, as shown in Figure 2.2. The first term now becomes a common mode charge and is therefore rejected, and Equation (2.7) becomes

$$V_o(t) = V_p(t) - V_n(t) = -2V_p\frac{\Delta C_s(t)}{C_f} \tag{2.8}$$

Note from (2.2) that the output voltage is proportional to the displacement of the resonant beam.

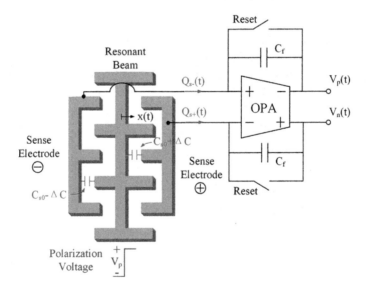

Figure 2.2   Differential capacitive sensing scheme based on discrete-time charge detection.

2.2 Front-end Amplifiers for MEMS Oscillators

As it can be seen later in Section 2.3, the core of the readout circuit for the silicon oscillating accelerometer (SOA) is the MEMS oscillator, which converts the external force or acceleration to frequency modulated output. Since they share similarities in front-end design specifications, namely, 1) large transimpedance to sustain oscillation, 2) wide bandwidth to avoid excess phase shift, and 3) low input referred current noise to improve SNR. In this section, the front-end amplifiers for MEMS oscillators will be discussed, which include single-stage resistive feedback TIA, two-stage resistive feedback TIA, capacitive feedback TIA, and CSA.

2.2.1 Single-Stage Resistive Feedback TIA

The concept of the single-stage resistive feedback TIA is shown in Figure 2.1(a). In practice, most MEMS processes, such as bulk micromachining, are not compatible with CMOS integrated circuit process. Two-chip solution is usually adopted, where MEMS device and CMOS ASIC or readout circuits are fabricated separately and integrated via wire bonding on a carrier, such as a PCB. Large parasitic capacitance exists at the input node of the TIA [3], as shown in Figure 2.3, where C_{p1} is the parasitic capacitance between the bonding pad and the substrate of the MEMS chip, and C_{p2} is the parasitic between ASIC chip bonding pads and its substrate. C_{p3} is the lumped parasitic capacitances on PCB, including that from the bonding wire. Besides, the MEMS structure capacitance, C_{s0}, also forms a part of the parasitic capacitance from the ASIC input to the ground since the proof mass is at a DC voltage, V_p [4]. Figure 2.4 shows the schematic of a resistive feedback TIA where C_{in} is the lumped input parasitic capacitance, R_f the feedback resistor, and C_c the stability compensation.

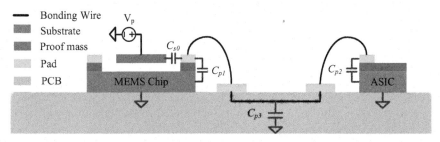

Figure 2.3 Integration diagram of MEMS and ASIC.

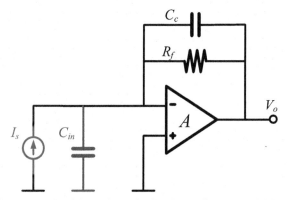

Figure 2.4   Schematic of the resistive feedback TIA front-end with input loading.

2.2.1.1 Stability and bandwidth

Assume that the OTA is a single-pole system with transfer function,

$$A(s) = \frac{A_0}{(1 + s/\omega_{p1})} \tag{2.9}$$

where A_0 is its open loop DC gain and ω_{p1} is its pole frequency.

The TIA has a shunt-shunt feedback topology with the feedback network transfer function, $\beta(s)$,

$$\beta(s) \approx -\frac{R_f C_c s + 1}{R_f} \tag{2.10}$$

and the open loop gain A'(s),

$$A'(s) = \frac{A_0}{(1 + s/\omega_{p1})} \cdot \frac{R_f}{1 + sR_f(C_c + C_{in})} \tag{2.11}$$

Thus, the loop gain is

$$L(s) = \beta(s)A'(s) = \frac{A_0(1 + sR_f C_c)}{(1 + s/\omega_{p1})[1 + sR_f(C_c + C_{in})]} \tag{2.12}$$

It can be seen that the input capacitance C_{in}, C_c, and R_f create a pole in addition to the pole from the op-amp, and furthermore, this additional pole is always lower than the zero frequency, which could have a serious impact on the stability of the TIA. This can be explained in Figure 2.5, where assuming ω_{p1} is the dominant pole. Since the two poles above 0 dB make the TIA unstable, the zero must be above 0 dB and should be made as close to the

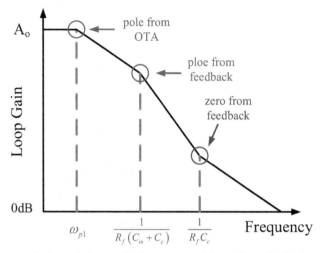

Figure 2.5 Loop gain magnitude response of the resistive feedback TIA.

second pole as possible so that the phase shift introduced by the second pole can be compensated. According to (2.12), to make the zero close to the second pole, a large compensation capacitor C_c is preferred. However, the large C_c will limit the bandwidth of the TIA for a given R_f. Thus, to make the TIA stable, its bandwidth has to be sacrificed.

For large loop gain, the transfer function of the TIA is approximately

$$R(s) \approx \frac{1}{\beta(s)} = \frac{-R_f}{R_f C_c s + 1} \tag{2.13}$$

and the bandwidth of the TIA is

$$BW = \frac{1}{2\pi R_f C_c} \tag{2.14}$$

In practice, considering the stability, C_c shall be in the same order of C_{in}, usually 1–10 pF. As an example, if a front-end TIA has a transimpedance of 10 MΩ and a C_c of 1 pF, for $C_{in} = C_c$, the bandwidth is limited to around 8 kHz. To increase the bandwidth by 10 times, the transimpedance or R_f has to be reduced by the same factor to 1 MΩ, which is a harsh trade-off between the gain and bandwidth.

2.2.1.2 Input-referred noise
The noise model of the TIA is shown in Figure 2.6(a): there are two noise sources in the TIA, namely, the current noise (I_n) from feedback resistor, R_f,

and the voltage noise (V_n) from the OTA. Since these two noises are uncorrelated, the TIA output voltage noise can be calculated by:

$$\bar{V}_{no}^2 = \bar{I}_n^2 \left(\frac{R_f}{R_f C_c s + 1} \right)^2 + \bar{V}_n^2 \left(1 + \frac{R_f C_{in} s}{R_f C_c s + 1} \right)^2 \qquad (2.15)$$

Since the input to the TIA is motion current from the MEMS resonator, the input referred current noise is usually used to evaluate the noise of the TIA. Dividing the output voltage noise by the front-end transimpedance, the input-referred current noise can be written as:

$$\bar{I}_{ni}^2 = \frac{\bar{V}_{no}^2}{\left(\frac{R_f}{R_f C_c s + 1} \right)^2} = \frac{4kT}{R_f} + \frac{8kT}{3g_m} \left[\frac{1 + sR_f \left(C_{in} + C_c \right)}{R_f} \right]^2 \qquad (2.16)$$

where $(v_n)^2 = 8\,kT/3g_m$ is assumed. From (2.16), it can be seen that the input-referred current noise contributed by R_f directly appears at the input and has a flat spectrum. The OTA-induced input current noise has a high-pass characteristic with the corner frequency at $[2\pi R_f(C_{in} + C_c)]^{-1}$, as plotted in Figure 2.6(b), where assuming that the equivalent input-referred current noise from the OTA is lower than that from R_f at a low frequency.

For the given C_c and C_{in}, the input-referred current noise can be reduced by increasing R_f and g_m. However, large R_f, though giving high trans-impedance, reduces the bandwidth, and large g_m may increase the power consumption. Thus, the low noise design needs to be balanced among gain, bandwidth, and power, as well as the stability.

From the above analysis, it can be seen that there are trade-offs in the resistive TIA. A large C_c alleviates the effect of C_{in} and improves the stability but sacrifices the bandwidth. Although a large R_f gives a high gain and reduces the input-referred current noise, it limits the bandwidth of the TIA and may result in a large chip area.

Despite the trade-offs, due to the simplicity of its circuit, and potentially small chip area and low power consumption, resistive TIA has been adopted as the front-end amplifier in MEMS oscillators and inertial sensors, such as in [4–6], to realize moderate gains ($\sim 10^6$ Ω) with reasonably wide band-widths ($\sim$MHz). One example is a MEMS oscillator fabricated in standard CMOS technology [6], as shown in Figure 2.7, whose oscillation frequency is 930 kHz with 2.1 MΩ motional resistance and consumes only 8.5 μW power. Its feedback transimpedance is implemented by a pair of PMOS transistor, M_3 and M_4 operating in the subthreshold region, as given in Figure 2.7.

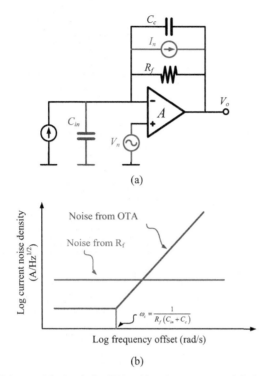

(a)

(b)

Figure 2.6 (a) Noise model of resistive TIA with noise sources and (b) input-referred current noise power spectrum plot.

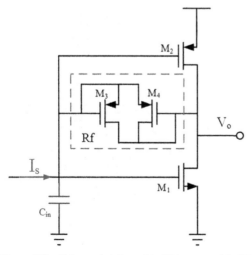

Figure 2.7 Schematic of on-chip TIA reported in [6].

Single-resistive TIA can also be employed in MEMS oscillator with two-chip solution but with very limited gain. Ref [4] reports a TIA in such an application with 20 kΩ transimpedance at 6 MHz, whose feedback resistor is also implemented by controlled impedance FET. Its input-referred current noise floor is 2.3 pA/Hz$^{1/2}$, which is dominated by the Op-amp noise and increased by input capacitance.

To overcome the harsh trade-offs among stability, GBW and noise performance in TIA, several improved architectures are proposed. The following sub-sections will review them in detail.

2.2.2 Two-Stage Resistive Feedback TIA

For MEMS resonators, the motion current to be detected is very low, in an order of nA. Thus, large transimpedance in tens of MΩ is usually required for the front-end amplifier, which demands a large R_f. According to the discussion in Section 2.2.1, although large R_f benefits the gain and input referred current noise, it reduces the bandwidth since the compensation capacitor, C_c, cannot be too small for a stable TIA. Depending on the implementation, a large R_f may also increase the chip area. To overcome these trade-offs, two-stage and T-network resistor TIA are proposed.

The typical topology of a two-stage TIA is shown in Figure 2.8 [7]. In the two-stage TIA, a second voltage gain stage is added to further amplify the output from the first stage TIA. The overall transimpedance R_{eq} is:

$$R_{eq} = R_f \left(1 + \frac{R_2}{R_1}\right) \tag{2.17}$$

In such a topology, the second stage enhances the total transimpedance by a factor of $(1+R_2/R_1)$. Therefore, for the same overall transimpedance as the

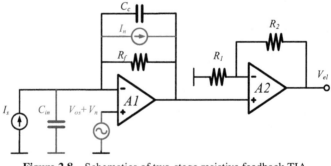

Figure 2.8   Schematics of two-stage resistive feedback TIA.

single-stage TIA, smaller R_f can be used in the first stage, which allows on-chip integration. Furthermore, the bandwidth is also benefited by the small R_f, which relaxes the trade-off between the high transimpedance and bandwidth. If assuming that the BW of the TIA is dominated by the first stage, for a given C_c and a same overall R_{eq}, the front-end bandwidth of the two-stage TIA is

$$BW_{2-stage} = \frac{1}{R_f C_c} = \left(1 + \frac{R_2}{R_1}\right) \frac{1}{R_{eq} C_c} \tag{2.18}$$

The bandwidth is increased by the same factor of $(1+R_2/R_1)$.

However, due to the small R_f, such a two-stage topology deteriorates the input referred current noise, which is given below, assuming that the first stage dominates,

$$\bar{I}_{ni}^2 = \left(1 + \frac{R_2}{R_1}\right) \frac{4kT}{R_{eq}} + \bar{V}_n^2 \left[\frac{1 + R_f (C_{in} + C_c) s}{R_f}\right]^2 \tag{2.19}$$

Compared with the input noise of the single-stage TIA in (2.16), the current noise from R_f is increased by $1+R_2/R_1$ times for the same overall transimpedance, R_{eq}. The OTA-induced input noise component at low frequency is also increased, but the high-pass noise portion remains the same since it is independent of R_f.

Besides, care should be taken when designing the two-stage TIA since the second gain boost stage not only amplifies the signal, but also the DC offset from the first stage, which could limit the dynamic range of the TIA, especially under low supply voltage. The power consumption of the two-stage resistive TIA is usually higher than that of the single-stage since two op-amps are required.

The above analysis can be summarized as follows: The two-stage resistive feedback TIA can realize the same transimpedance as that of single-stage TIA with a feedback resistor (R_f) that is $1+(R_2/R_1)$ times smaller. Thus, its bandwidth is also broadened by the same factor. In addition, small R_f facilitates the on-chip integration. However, this is done at the penalties of higher input referred current noise. The stability issue caused by the input parasitic capacitance still exists and needs to be carefully considered during the design, as it ultimately limits the bandwidth improvement.

One implementation of the two-stage TIA for a MEMS oscillator is shown in Figure 2.9 [8]. A_{mp2} boosts the overall transimpedance and thus a smaller R_f can be employed to broaden its bandwidth. The TIA is fabricated in a $0.18-\mu m$ CMOS process, achieving 110 dBΩ transimpedance and 60 MHz

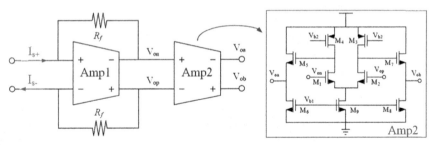

Figure 2.9 Schematic of two-stage TIA reported in [8].

bandwidth. Its input current noise is 2.5 pA/Hz$^{1/2}$ at 17.6 MHz with 5.9 mW power consumption. In this design, the input referred current noise is dominated by the amplifier noise. The main benefit of the two-stage topology here is the increased bandwidth, while maintaining relatively high overall transimpedance.

2.2.3 T-Network Resistive Feedback TIA

Another technique to reduce R_f while achieving high transimpedance and stability is to use a T-network resistor in the feedback of the TIA, as shown in Figure 2.10 [9].

With T-network resistor, large resistance can be realized with three small resistors connected in a T-network. The overall transimpedance can be expressed as

$$R_{eq} = R_f \left(1 + \frac{R_2}{R_1} \right) + R_2 \tag{2.20}$$

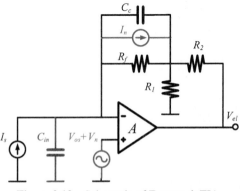

Figure 2.10 Schematic of T-network TIA.

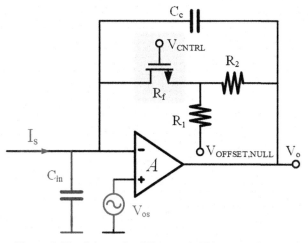

Figure 2.11 Schematic of T-network TIA reported in [9].

where it can be seen that R_f is enlarged by the ratio of $(1+R_2/R_1)$. In other words, to achieve the same R_{eq}, the R_f in T-network TIA can be $(1+R_2/R_1)$ times smaller than that in the single-stage resistive TIA. Since the compensation capacitor, C_c, is shunt only with R_f, the bandwidth and input referred current noise are all identical with the two-stage TIA. This TIA is more compact as only one op-amp is required. However, the OTA DC offset is also amplified by the feedback network, thus needs to be minimized.

Figure 2.11 shows a T-network resistive TIA reported for a MEMS oscillator in gyroscope [9]. The main transimpedance R_f is implemented by a controlled FET, and a regulation voltage is applied via R_1 to trim the DC offset. Fabricated in 0.6-μm CMOS process, it achieves 1.6 MΩ transimpedance and 200 kHz bandwidth. The input-referred current noise is 88 fA/Hz$^{1/2}$ at 15 kHz. It consumes 400 μW power under a $\pm$1.5 V supply. The noise performance is limited by the reduced feedback resistor R_f.

2.2.4 Charge-Sensing Amplifier (CSA)

The two-stage and T-network resistive feedback TIAs discussed above relax the trade-off between transimpedance gain and its bandwidth. However, this is achieved at a penalty of a much higher input-referred noise. An alternative is the CSA, which provides superior noise performance and inherently stable [10]. Figure 2.12 shows its schematic. Although the circuit is similar to the resistive TIA, in CSA, the transimpedance gain is set by C_f, and R_b only provides a DC feedback.

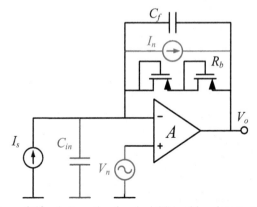

Figure 2.12   Schematic of typical CSA with noise sources.

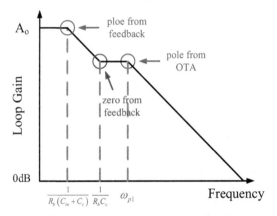

Figure 2.13   Loop gain magnitude response of the CSA.

R_b is usually implemented by a pseudo resistor (i.e. two or more seriously connected PMOS transistors biased in subthreshold region), as shown in Figure 2.12. Since the pseudo resistor can have very high resistance, in an order of $10^{12}\Omega$ [11], the resultant pole with the C_f and C_{in} is much lower than the dominant pole from the OTA. Therefore, although CSA has the same loop gain equation as in (2.12), its magnitude response is different from the resistive TIA in Figure 2.5. Figure 2.13 shows magnitude response of the loop gain of the CSA. Since the zero from feedback network cancels the phase shift from the pole of the OTA (ω_{p1}), the stability of CSA is inherently guaranteed, even at the presence of the large input parasitic, C_{in}.

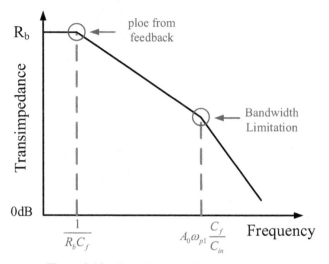

Figure 2.14 Close loop magnitude plot of CSA.

Using Miller equivalence, the closed-loop transfer function of CSA can be derived as:

$$R(s) = \frac{R_b}{R_b \left(C_f + C_{in}/A\left(s\right) \right) s + 1} \approx \frac{R_b}{R_b C_f s \left(1 + \frac{C_{in}}{C_f} \frac{1}{A_0 \omega_{p1}} s \right) + 1}$$

(2.21)

As shown in Figure 2.14, it shows a first-order low-pass response until the second pole at $(A_0 \omega_{p1})*(C_f/C_{in})$, which indicates its closed-loop bandwidth is C_{in}/C_f times lower than the GBW of the OTA. Since the first pole $1/(R_b * C_f)$ is near DC, the CSA behaves like an integrator, which integrates the input motion current onto the feedback capacitance, C_f, and produces an output voltage. Due to the integration, the output phase is 90° lagging the input, thus the output of CSA is proportional to the resonator's displacement. In the case of the MEMS oscillator, another +90° phase shift is needed in order to satisfy the oscillation phase condition.

For noise analysis, (2.16) is also valid for CSA after replacing R_f and C_c with R_b and C_f. In CSA, the voltage noise from OTA still has a high-pass characteristic and usually dominates the overall noise floor, since the extremely large pseudo resistor contributes negligible current noise.

In summary, CSA is inherently stable and has superior noise performance since the feedback capacitor is noiseless and the large pseudo resistor has negligible noise current. However, the existence of input capacitive loading, C_{in}, still limits its bandwidth and increases the input referred current noise originated from the OTA noise. Moreover, since the output of CSA is in phase with the resonator's displacement, a phase shifter is required to form an oscillator.

2.2.5 Capacitive Feedback TIA

To utilize the low noise characteristic of CSA, capacitive feedback TIAs with oscillation velocity output have been proposed. One capacitive feedback TIA reported in [12] adopts same circuit topology from [13], but is optimized for MEMS oscillator application.

Figure 2.15 shows a single-ended schematic of the capacitive feedback TIA. It is essentially a current amplifier with capacitive feedback, followed by an I-V converter. The capacitive feedback network is noise free and therefore has no noise contribution to the input of the TIA, while the I-V converter is realized by R_f whose noise is attenuated by the gain of the preceding current amplifier. This allows a smaller R_f, which benefits the bandwidth without introducing high input current noise. The current amplifier, based on shunt-series feedback, consists of an op-amp, A, M_1, C_1 and C_2. The current gain can be written as

$$\frac{I_0}{I_s} = \frac{1}{\beta} = \left(1 + \frac{C_2}{C_1}\right) \qquad (2.22)$$

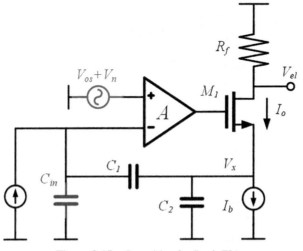

Figure 2.15   Capacitive feedback TIA.

where β is the feedback factor and equal to $C_1/(C_1 + C_2)$. Here its loop gain much greater than one is assumed. The current output is converted to voltage through R_f, and the overall transimpedance of the TIA is:

$$R_{eq} = \frac{I_0 R_f}{I_s} = \left(1 + \frac{C_2}{C_1}\right) R_f \qquad (2.23)$$

It can be seen from Figure 2.15 that there are two open-loop poles in the current amplifier, at the 3 dB frequency of the op-amp, ω_{3dB}, and the source of M_1, $\sim g_{m1}/C_2$, respectively. The bandwidth of the closed-loop current amplifier depends on which of these two poles is dominant. For $\omega_{3dB} >> g_{m1}/C_2$, the current amplifier 3 dB bandwidth can be found from small-signal analysis,

$$f_{3db,TIA} = \frac{g_{m1}}{2\pi C_2} \left(\frac{A_0 C_1}{C_1 + C_{in}}\right) \qquad (2.24)$$

and for $\omega_{3dB} << g_{m1}/C_2$, the bandwidth is:

$$f_{3db,TIA} = \frac{\omega_{3dB}}{2\pi} \left(\frac{A_0 C_1}{C_1 + C_{in}}\right) \qquad (2.25)$$

When the pole at the output of the TIA is considered, the overall transfer function of the TIA is

$$\frac{V_o(s)}{I_s(s)} = \frac{I_0(s)}{I_s(s)} \frac{R_f}{1 + sR_f C_L} \qquad (2.26)$$

where there are three poles. However, the pole associated with R_f is outside of feedback loop and thus it doesn't cause stability problem but should be higher than the bandwidth of the TIA.

The input referred current noise is given by:

$$\bar{I}_{ni}^2 = \frac{4kT}{R_f (1 + C_2/C_1)^2} + \bar{V}_n^2 \left[s (C_1 + C_{in})\right]^2 \qquad (2.27)$$

where the first term is the current noise from R_f and the second term the noise from the op-amp. It can be seen that current noise from R_f is attenuated by the current amplifier gain, $(1+C_2/C_1)$, which implies that for the same R_f, lower input referred noise can be achieved, compared with the resistive feedback TIAs. The op-amp noise is differentiated or high-pass filtered before referring to the input of the TIA.

Although capacitive feedback is noise free, it lacks DC feedback and the input is virtually floating. Thus, additional circuit is needed to provide the DC

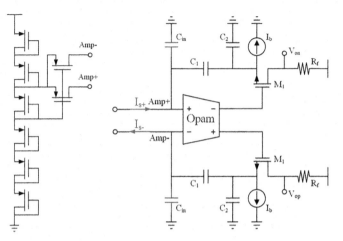

Figure 2.16   Schematic of capacitive feedback TIA and its bias circuit reported in [12].

bias for the op-amp. Furthermore, since the TIA is DC open-loop, its input DC offset is amplified by the open-loop gain, A_0, and causes large output offset, which may limit the dynamic range of TIA.

The practical implementation of the above-discussed TIA in differential topology is shown in Figure 2.16 [12]. A stack of diode-connected PMOS transistors operating in the subthreshold region is employed to provide the DC input bias. However, the DC offset problem still exists due to the lack of DC feedback. The TIA demonstrates a 56 MΩ transimpedance and 1.8 MHz bandwidth. The input current noise density at 100 kHz is 65 fA/Hz$^{1/2}$ with 436 μW power. In this work, the frequency response of the TIA is designed for a maximally flat response.

2.3 Front-end Amplifier for MEMS SOA

2.3.1 Concept of MEMS SOA and its Front-end

Figure 2.17 presents a schematic of MEMS SOA and its readout circuit as well as its equivalent model. MEMS SOA usually consists of a proof mass and two resonators, which are embedded into two oscillation circuits in the readout circuit. When subjected to an external acceleration, the frequencies of the two resonators shift differentially, and the frequency difference represents the input acceleration.

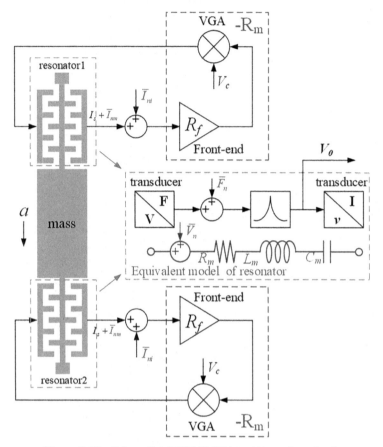

Figure 2.17 Schematic of MEMS SOA and its readout circuit.

The electrical equivalent model of MEMS resonator is also given in Figure 2.17; its behaviour equals to a series R-L-C network, whose parameters are given as [14]:

$$R_m = \frac{m\omega_0}{QK_{vf}K_{vi}} \quad L_m = \frac{m}{K_{vf}K_{vi}} \quad C_m = \frac{K_{vf}K_{vi}}{m\omega_0^2} \quad (2.28)$$

in which K_{vf} and K_{vi} represent the capacitive transducer gain from the drive voltage to the drive force and from the resonator beam velocity to the motion current, respectively. V_c is the output from automatic amplitude control loop (not shown in the figure) that controls the overall loop gain of the oscillator via VGA and makes it equal to $-R_m$ to compensate the energy loss in the MEMS resonator. In the readout circuit, the front-end amplifier senses the motion current and converts it to a voltage. The resonator's mechanical Brownian noise can be modelled as a drive force noise source and equivalent to the voltage noise source of R_m with power density as [15].

$$\bar{V}_{nm}^2 = \frac{4kTm\omega_0}{QK_{vf}K_{vi}} \tag{2.29}$$

To start and sustain the oscillation with MEMS resonator, the front-end must provide a transimpedance (R_f) with its magnitude larger than resonator's motional resistance R_m (in an order of MΩ). To tolerate process variation and accelerate the start-up process, R_f is usually designed to be larger than R_m. On the other hand, front-end shall also provide a near $0°$ phase shift to guarantee that the resonator oscillates at its natural frequency, ω_0, with the highest Q factor.

As seen in Figure 2.17, the current noise at the input of the front-end are the input referred current noise from the TIA and the Brownian noise from the MEMS resonator. According to the phase noise model in [16–18], these white current noises will be folded into the $1/f^2$ phase noise region in the phase noise spectrum, while the thermal noise from the front-end TIA alone determines phase floor, as illustrated in Figure 2.18. The corner frequency in the phase noise spectrum is denoted, ω_c. In the design, the front-end current noise floor shall be lower than the Brownian noise so that the $1/f^2$ phase noise of the MEMS oscillator is dominated by the MEMS Brownian noise. Besides, the front-end current noise should be as low as possible to move ω_c to a higher frequency. This will increase the bandwidth of the SOA while maintaining its resolution, since when converted to frequency noise spectrum, $1/f^2$ phase noise becomes white frequency noise and determines the frequency noise floor, while the white phase noise floor becomes the frequency noise with a slope of f^2.

Besides, to sustain the oscillation, the front-end amplifier must provide an output that is in phase with the velocity of the resonator beam, to satisfy the phase condition for oscillation. An automatic amplitude control (AAC) circuit is usually employed to stabilize the oscillation amplitude and hence the oscillator loop gain. The amplitude information can be extracted from the output of the front-end amplifier. On the other hand, the amplitude-stiffness (A-S) effect induced frequency variation could affect the scale factor linearity of the SOA [19]. The A-S effect is directly related to the MEMS resonator beam displacement amplitude. Thus, it is desirable to stabilize the displacement amplitude. To extract the displacement amplitude information

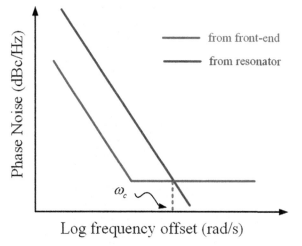

Figure 2.18   Phase noise plot of MEMS Oscillator.

for AAC, the output proportional to the displacement should be available from the front-end amplifier.

2.3.2 Continuous-Time Integrator-Differentiator-Based TIA

Figure 2.19 shows an integrator and differentiator-based TIA [19]. It is a two-stage TIA with bandpass characteristic. The first stage is an integrator that integrates the motion current onto C_1, and its output voltage is proportional to the displacement of the MEMS resonator beam due to the 90^0 phase shift from the integrator. R_b is a PMOS-based pseudo resistor network providing DC feedback path to Opa1. The lower cut-off frequency, ω_L, from R_b and C_1 is designed to be much lower than ω_0 to avoid the extra phase shift at the signal frequency ($\sim$25 kHz). The second stage can be considered as a differentiator that introduces a 90° leading phase shift and cancels the 90° lagging phase from the first integrator stage. Thus, the output of overall TIA is proportional to the velocity of the resonator beam, satisfying the phase condition for the oscillator. C_c is a stability compensation capacitor for Opa2 and, together with R_f, sets the high-pass corner frequency of the differentiator, ω_H, which is also the upper cut-off frequency of the bandpass TIA. The overall transfer function can be expressed as:

$$H(s) = \frac{R_b R_f C_2 s}{(R_b C_1 s + 1)(R_f C_c s + 1)} \qquad (2.30)$$

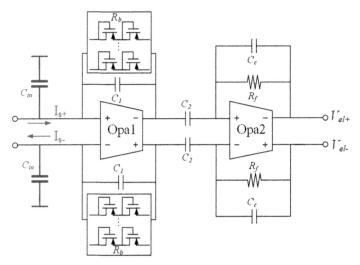

Figure 2.19   Integrator and differentiator-based TIA.

with cut-off frequencies being

$$\omega_L = \frac{1}{R_b C_1}, \quad \omega_H = \frac{1}{R_f C_c} \tag{2.31}$$

The overall magnitude response is shown in Figure 2.20. With properly chosen ω_L and ω_H, the TIA has a band-pass characteristic. Its mid-band transimpedance is:

$$R_{eq} = R_f \frac{C_2}{C_1} \tag{2.32}$$

It can be seen that a large transimpedance can be achieved by a small value of R_f and the proper ratio of C_2/C_1. The expression in (2.32) is similar to the capacitive TIA in Section 2.2.5. As explained in 2.2.4, the extra pole introduced by C_{in} at the input of the first stage can be inherently compensated by the zero in feedback path. Therefore, the stability of Opa1 is independent of C_{in}.

However, if we have a close look at the transfer function of the first stage, as given in (2.21), C_{in} forms the feedback around Opa1 with C_f. Therefore, the maximum frequency of integrator is limited by the closed-loop bandwidth of Opa1 and given below

$$\omega_{I,\mathrm{ma}} = \frac{C_1}{C_{in}} GBW_1 \tag{2.33}$$

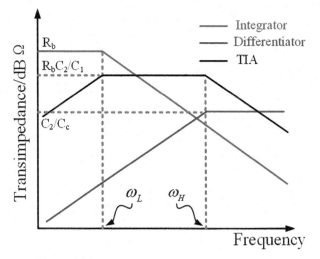

Figure 2.20 Magnitude response of bandpass TIA.

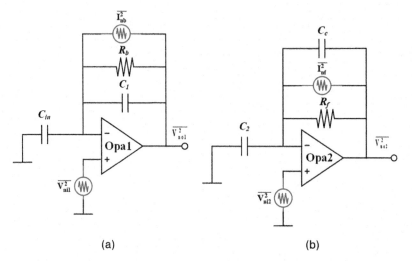

Figure 2.21 Noise model of TIA (a) first stage and (b) second stage.

where GBW_1 is the gain-bandwidth product of Opa1. Apparently, the bandwidth limitation caused by C_{in} still exists in this topology.

Figure 2.21 presents its noise analytic model, there are four noise sources from resistors and Op-amps. The noise behaviours of $\bar{V}_{ni1}^2$ and $\bar{I}_{nb}^2$ are identical with what described in (2.16), the total input referred current noise can be calculated by adding the noise contribution from the second stage.

$$\bar{V}_{no2}^2 = \bar{I}_{nf}^2 \left(\frac{R_f}{R_f C_c s + 1} \right)^2 + \bar{V}_{ni2}^2 \left(1 + \frac{R_f C_2 s}{R_f C_c s + 1} \right)^2 \quad (2.34)$$

The total input referred current noise power density can be computed as:

$$\bar{I}_{ni}^2 = \bar{I}_{nb}^2 \left[\frac{R_b C_1 s}{(s/\omega_L + 1)(s/\omega_H + 1)} \right]^2 + \bar{I}_{nf}^2 \left(\frac{C_1/C_2}{s/\omega_H + 1} \right)^2$$
$$+ \bar{V}_{ni1}^2 \left(1 + \frac{R_b C_{in} s}{s/\omega_L + 1} \right)^2 \left(\frac{C_1 s}{s/\omega_H + 1} \right)^2 + \bar{V}_{ni2}^2 \left(\frac{C_1 s}{s/\omega_H + 1} \right)^2 \quad (2.35)$$

For a given oscillation frequency, ω_0, it satisfies $\omega_L \leq \omega_0 \leq \omega_H$. Therefore, its in-band input referred current noise can be simplified to:

$$\bar{I}_{ni}^2 = \bar{I}_{nb}^2 + \bar{I}_{nf}^2 \left(\frac{C_1}{C_2} \right)^2 + \bar{V}_{ni1}^2 \left[s \left(C_1 + C_{in} \right) \right]^2 + \bar{V}_{ni2}^2 \left(s C_1 \right)^2 \quad (2.36)$$

The current noise (I_{nb}) from R_b directly appears at the input of the TIA. The current noise (I_{nf}) from R_f is attenuated by capacitor ration C_2/C_1 before referring to the input. The noises from Opa1 and Opa2 (V_{ni1} and V_{ni2}) are high-pass shaped, and C_{in} increases the noise floor in the first stage.

The input current noise floor is usually dominated by the second term, as long as the below criteria are satisfied.

$$R_b > R_f \left(\frac{C_2}{C_1} \right)^2 = R_{eq} \frac{C_2}{C_1} \quad (2.37)$$

$$\bar{V}_{ni1}^2 < \frac{4kT/R_f}{\left[\omega_0 C_2 \left(1 + C_{in}/C_1 \right) \right]^2} \quad (2.38)$$

$$\bar{V}_{ni2}^2 < \frac{4kT/R_f}{\left(\omega_0 C_2 \right)^2} \quad (2.39)$$

Equation (2.37) can be satisfied by adopting a pseudo resistor for R_b, which can easily achieve $> 10^9 \Omega$. Equation (2.38) and (2.39) are the noise design criteria for Opa1 and Opa2, from which the power budget can be allocated to two op-amps. More current should be distributed to Opa1 due to the existence of C_{in}.

In summary, this TIA has several advantages over other mentioned topologies: 1) large transimpedance can be realized with small R_f and a proper ratio of C_2/C_1, its transimpedance and bandwidth can be independently designed; 2) the noise contributed from two Opamps are both high-pass shaped by the integrator gain in first stage, the needed power for low noise performance is therefore greatly reduced; 3) the dominant noise from R_f is attenuated by the ratio of C_2/C_1 to further reduce the input referred current noise floor; 4) the displacement and velocity of MEMS resonator are available for amplitude control and feedback drive force generation; 5) the phase margin deterioration arisen from C_{in} is inherently compensated by a large value of R_b; 6) R_b and R_f provide DC feedback path for two Opamps and the differentiator input is AC coupled; DC offset is no longer a big concern.

The integrator and differentiator-based TIA in [19] for a MEMS SOA achieves a transimpedance of 45 MΩ with a feedback resistor of 1 MΩ and a C_2/C_1 ratio of 45. It has a bandwidth of 0.5 Hz–350 kHz, as given in Figure 2.22(a). The output voltage noise floor is 300 nV/Hz$^{1/2}$(see Figure 2.22(b)), corresponding to 6.6 fA/Hz$^{1/2}$ input referred current noise floor with 583 μW power.

By adopting this front-end, the SOA achieves 2 μg/Hz$^{1/2}$ noise floor, which is dominated by the mechanical noise of MEMS resonator. It also achieves 0.6 μg bias instability, as presented in Figure 2.23.

This kind of front-end amplifier has also been employed in MEMS oscillators [20, 21]. Both of them achieve 8 MΩ transimpedance and 1.2 MHz bandwidth with <25 fA/Hz$^{1/2}$ input referred current noise floor.

2.3.3 Discrete-Time Integrator-Differentiator-Based Amplifier

The front-ends described so far are operating continuously in time domain, i.e., they continuously sense the signal (motion current) induced by the acceleration or proof mass displacement, referring to as continuous-time (CT) front-ends. Discrete-time (DT) front-ends, on the other hand, have also been employed for capacitive sensing in MEMS sensors, such as those in SOAs [22, 23] and in capacitive accelerometers [25–29]. The concept of a DT front-end amplifier shown in Figure 2.24 is similar to the CSA in Figure 2.2 in Section 2.1.

A periodic step voltage, V_p, is applied to the common plate of the sense capacitors. A charge change ΔQ is therefore generated from capacitance change of the sense electrode, ΔC, induced by the proof mass displacement. The ΔQ is then transferred to the feedback capacitor, C_f, and converted to an output voltage. The charge on C_f is reset before each sense period. The DT front-end effectively senses the capacitor change (ΔC) and converts it to a voltage. It is sometimes referred to as a C-to-V converter. The conversion

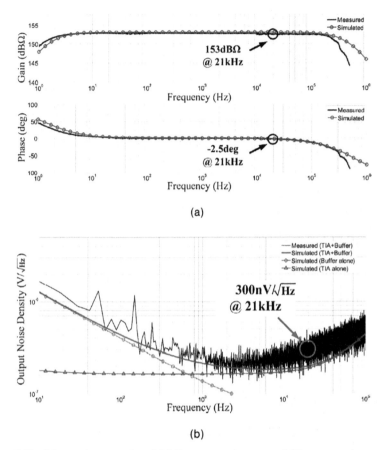

Figure 2.22   Measurement results of (a) frequency response and (b) output noise spectrum in [19] (Courtesy of IEEE press).

gain is dependent on V_P and C_f and given below:

$$V_o(t) = -\frac{2V_p}{C_f}\Delta C(t) \qquad (2.40)$$

The major noise sources in the DT CSA are kT/C noise, switch charge injection, DC offset arising from sense capacitance mismatch, and the noise from the OPA. Correlated double sampling (CDS) technique is usually employed to attenuate the DC offset, flicker noise, charge injection and the thermal noise at low frequency [24–27]. CDS is based on the fact that the low frequency noises are correlated between two adjacent samples, which can be cancelled

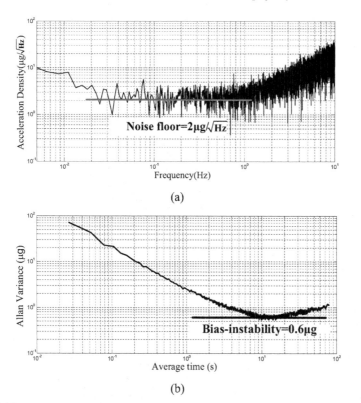

(a)

(b)

Figure 2.23 Measurement results of (a) acceleration output spectrum and (b) Allan variance in [19] (Courtesy of IEEE press).

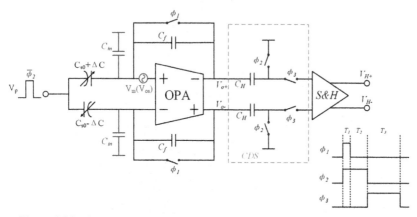

Figure 2.24 Schematic of a typical switched-cap front-end with CDS operation.

by taking the difference of the two adjacent samples. In Figure 2.24, C_H stores the low frequency errors of OPA and subtract them from the output in the subsequent phase.

The input-referred current noise power density of DT CSA can be expressed as [24]:

$$\bar{I}_{ni}^2 = (2\pi f)^2 \left[\frac{kTC_f}{f_s/2} + \frac{f_B}{f_s/2} \bar{V}_{ni}^2 \left(C_{s0} + C_{in} + C_f \right)^2 \right] \qquad (2.41)$$

where V_{ni} is the input referred noise of the OPA and, f_B is the equivalent noise bandwidth of the OPA, f_s is the sampling frequency. The first term is the noise from the reset switch, which is sampled on C_f, and the second term is due to the thermal noise from the OPA. The OPA contributed input-referred current noise is similar with that in (2.16), but multiplied by a noise folding ratio $2f_B/f_s$ arisen from CDS and S&H.

To guarantee the settling accuracy, the bandwidth of the OTA must be much larger than f_s. Large bandwidth results in high total in-band noise and noise folding. To maintain the same noise performance, intrinsic noise from key transistors need to be reduced with increased power consumption. Since both low noise and wide bandwidth demand high power consumption, the DT CSAs are usually more power hungry than its CT counterpart.

Despite of its high-power consumption, one of the advantages of the DT CSA is that the feedback drive and sense operations can be carried out in two different periods to avoid the drive feedthrough interference via the parasitic capacitance between the input and output terminals of the resonator [28–30]. As shown in Figure 2.25, the parasitic capacitance, C_f, arising from package and inner traces of MEMS resonator couples the drive signal to the front-end input node. The feedthrough interference, i_{s2}, corrupts the sensing current signal, i_{s1}, from MEMS resonator, adding an additional phase shift in the oscillation loop [30].

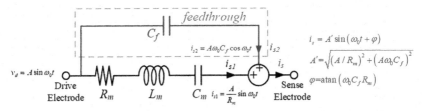

Figure 2.25   MEMS resonator electrical equivalent model with feedthrough effect.

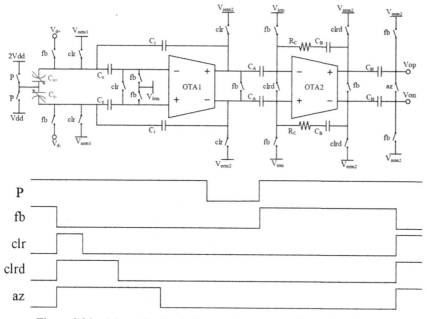

Figure 2.26 Schematic of switched-cap integrator and its timing diagram.

A switched-cap front-end with the integrator and differentiator scheme is reported in [23]. Its first stage is a charge integrator followed by another gain stage. A 5 MHz sampling clock is used to oversample the charge variation signal at 135 kHz. Its operation can be divided into four phases, namely, clear (clr), autozero (az), sense (~P) and drive (fb), as shown in Figure 2.26.

In *clear* phase, the input nodes of charge amplifier are biased to common mode voltage V_{icm} and the charge across C_i is erased. In *autozero* phase, the amplifier's DC offset and 1/f noise are stored on the left plate of C_H. During the *sense* phase, the polarization voltage, V_p, applied to the MEMS resonator, steps down from $2V_{DD}$ to V_{DD}, and the charge proportional to $(C_{s+} - C_{s-})$ is transferred to C_i and the offset and 1/f noise stored on previous phase is now subtracted from the charge amplifier output. During the *drive* phase, a pair of differential drive voltage V_d is applied on the same electrodes for charge sensing, meanwhile, the polarization voltage recovers to $2V_{DD}$. A pair of capacitors C_c isolates the drive voltage from amplifier input nodes to prevent it from saturation.

Since the output of the front-end is proportional to the displacement of the resonator, a differentiator is added after the front-end to generate drive voltage

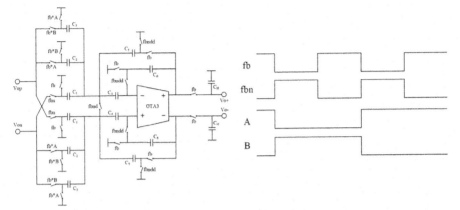

Figure 2.27 Schematic of switched-cap differentiator with its clock diagram.

that is proportional to the velocity of the MEMS resonator. Figure 2.27 shows its schematic. The differentiation function is realized by charge redistribution among two capacitors. C_1 stores the present signal while C_2 and C_3 alternatively store the delayed signal. CDS is also employed to cancel the low frequency noise. C_4 stores *ota3*'s low frequency noise, including DC offset, when *fbn* is high and cancels it in the next phase when *fb* is high.

The MEMS SOA achieves 20 μg/Hz$^{1/2}$ noise floor and 4 μg bias instability with 23 mW power consumption.

2.3.4 Front-end Based on Passive Charge Sensing

The charge sensing or C-to-V conversion can also be implemented with only passive components, which is the case in Pierce oscillators [31–33], as shown in Figure 2.28, and Pierce oscillator-based MEMS SOA [34]. Transistor, M_1, with C_1 and C_2, generate a negative resistance that matches the motional resistance, R_m, of the MEMS resonator, and thus form an oscillator. C_{p1} and C_{p2} are parasitic capacitances from package, which can be lumped into C_1 and C_2, respectively.

In Figure 2.27(a), C_1 effectively integrates the motion current and generates a voltage, V_{c1}, that is proportional to resonator's displacement. V_{c1} is subsequently converted to drain current of M_1 through its transconductance, g_m, and the drain current is integrated on C_2 and generates the drive voltage,

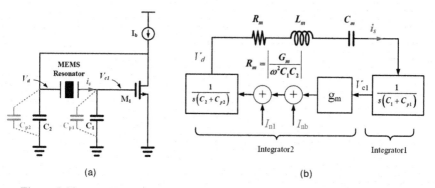

Figure 2.28   (a) schematic of Pierce oscillator and (b) its equivalent circuit model.

V_d, for the MEMS resonator. Unlike the two-stage TIA, the feedback drive voltage is generated by two integrations and thus satisfies the phase condition.

In Pierce oscillator, the only noise source from the circuit is the noise in the drain current of M_1, which consists the current noises from both M_1, I_{n1}, and bias current source transistor, I_{nb}. When referring to the motion current sense node, its input referred current noise power density can be calculated as

$$\bar{I}_{ni}^2 = \left(\frac{8kT}{3g_m} + \frac{8kTg_{mb}}{3g_m^2} \right) [s(C_1 + C_{p1})]^2 \qquad (2.42)$$

where g_{mb} is the transconductance of current bias transistor. It can be seen although the first passive integrator is noise less, the overall input referred current noise power remains the same as the charge amplifier in Section 2.2.4 Although large g_m can reduce the noise, it is constrained by the value that matches the motional resistance, R_m, to sustain the oscillation, as given in (2.43). The ultimate input referred current noise floor of front-end amplifier also depends on the parasitic capacitances.

$$R_m = \left| \frac{g_m}{\omega_0^2 (C_1 + C_{p1})(C_2 + C_{p2})} \right| \qquad (2.43)$$

The main advantage of the Pierce oscillator-based front-end is the low power consumption since only one active device is required, which is attractive to consumer applications. However, its limitation is also clear from (2.42). To reduce the noise of the front-end, small g_m is preferred. However, g_m cannot

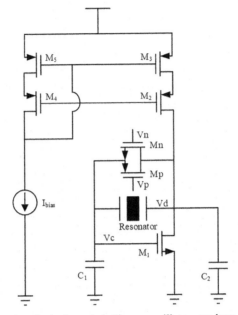

Figure 2.29   Schematic of single-ended Pierce oscillator readout circuit in a MEMS SOA [34].

be arbitrarily chosen for a given R_m. On the other hand, minimum C_1, and C_2 are limited by parasitic capacitance. As a result, its noise floor cannot be optimized as desired.

A single-ended Pierce oscillator-based readout circuit for MEMS SOA was demonstrated in [34], as shown in Figure 2.29. Capacitors C_1 and C_2 are chosen to be 3.5 pF. To minimize the parasitic capacitance, the ASIC die is glued on top of the wafer-level packaged MEMS chip. A compact amplitude-limiting network formed by M_n and M_p is employed to stabilize the oscillation amplitude without adding extra power. The Pierce oscillator core is biased with 4 μA current under a 1.8 V supply. It achieves 360 μg/Hz$^{1/2}$ noise floor with only 21.6 μW power.

Another Pierce oscillator-based MEMS SOA was later reported in [35], as shown in Figure 2.30. Differential topology is employed to suppress common mode noise from the bias circuit, and the amplitude control is realized by regulating the transconductance (g_m) of the OTA via its bias current. It achieves 10 μg/Hz$^{1/2}$ noise floor and 4-μg bias instability while consuming only 27 μW power.

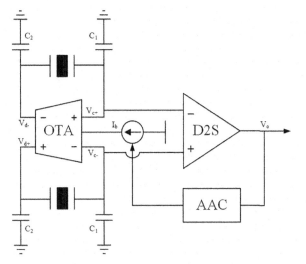

Figure 2.30 Schematic of differential Pierce oscillator-based readout circuit in MEMS SOA [35].

2.4 Summary

In this chapter, integrated CMOS front-end amplifiers for MEMS oscillators and SOAs have been reviewed and described. The front-end amplifier in the MEMS SOA, usually a transimpedance amplifier or TIA, senses the motion current from MEMS sensor induced by the external acceleration via the tiny electrode capacitance change. The key parameters of the CMOS front-end amplifiers include transimpedance gain, noise, bandwidth, stability, and power consumption. Trade-offs among these parameters are often needed.

Single-stage resistive TIA has a simple circuit, but there are trade-offs among its transimpedance, bandwidth and chip area. The stability of the TIA is sensitive to input parasitic capacitance. Two-stage and T-network resistive TIAs can achieve high transimpedance with a small resistor, thus relaxes the above trade-offs. However, this is at the cost of high input-referred noise and dealing with the DC offset.

Charge-sensing amplifier (CSA), where the transimpedance is provided by feedback capacitance, can achieve high transimpedance and easily satisfy the stability requirement at the presence of parasitics at the TIA input. The input-referred current noise is dominated by the noise from OTA since the DC bias network has considerably low current noise. There is still a trade-off between the transimpedance and bandwidth, but can be alleviated by increasing the bandwidth of the OTA. However, the CSA output is proportional to the displacement of MEMS resonator. Additional stage is needed to provide 90° phase shift for the feedback drive signal and satisfy the oscillation phase condition.

The capacitive feedback TIA and integrator-differentiator-based TIAs are two topologies whose outputs are proportional to the resonator beam's velocity, and therefore can be directly fed back to drive the resonator in an oscillator. Both of them immune the effect of large parasitic capacitance at the input on the stability. The integrator-differentiator based TIA can provide displacement (output of the first stage) and velocity (output of the TIA) outputs simultaneously, which is particularly suitable for the SOAs with displacement control strategy.

Passive charge sensing-based front-end, as in Pierce oscillator, uses a capacitor to integrate the motion current, followed by a G_m cell and another integrator to generate a negative R_m and sustain the oscillation. It has the simplest circuit and hence a very low power consumption. However, one drawback is that both its transimpedance and noise floor are sensitive to the parasitic capacitance at the interconnections of the MEMS transducer

Table 2.1   Comparison of front-end solutions for MEMS applications

Paper	Topology	Application	Gain	Current Bandwidth	Noise Floor	Power
[4]	Single-resistor	oscillator in gyroscope	20 kΩ	6 MHz	2.3 pA/Hz$^{1/2}$	-
[6]	Single-resistor	oscillator	930 kΩ	2.1 MHz	-	8.5 μW
[9]	T-network	oscillator in gyroscope	1.6 MΩ	230 kHz	88 fA/Hz$^{1/2}$	400 μW
[8]	Two-stage	oscillator	316 kΩ	60 MHz	2.5 pA/Hz$^{1/2}$	5.9 mW
[12]	Capacitive feedback	oscillator	56 MΩ	1.8 MHz	65 fA/Hz$^{1/2}$	436 μW
[20]	Int-Diff	oscillator	8 MΩ	1.2 MHz	25 fA/Hz$^{1/2}$	50 μW
[19]	Int-Diff	SOA	45 MΩ	350 kHz	6.6 fA/Hz$^{1/2}$	583 μW
[33]	Gm-based	oscillator	100 MΩ	3 MHz	25 fA/Hz$^{1/2}$	1.2 mW
[21]	Int-Diff	oscillator	8 MΩ	1.2 MHz	25 fA/Hz$^{1/2}$	150 μW
[32]	Gm-based	oscillator	4 MΩ	24 MHz	166 fA/Hz$^{1/2}$	-

and readout circuit. Besides, its overall noise floor cannot be improved by consuming more current in front-end, since the value of g_m is not arbitrary, but decided by the resistance of MEMS resonator.

Switched-cap charge amplifier transfers the charge on sense electrode to a capacitor in discrete-time. It can perform drive and sense operations alternatively in two different time slots and therefore avoids the feedthrough between the drive and sense nodes. It is well suited for switched-capacitor-based discrete-time readout circuit with CDS noise cancellation. Due to its sampling nature and the setting time requirement, it consumes more power than its CT counterparts.

The chapter concludes with a comparison among different types of front-end amplifiers for MEMS resonators. Table 2.1 summarises the performance of published typical front-end amplifiers.

References

[1] N. Yazdi, F. Ayazi and K. Najafi., "Micromachined inertial sensors". *Proceedings of the IEEE*, 1998. 86(8): pp. 1640–1659.

[2] N. Barbour and G. Schmidt., "Inertial sensor technology trends". *IEEE Sensors Journal*, 2001. 1(4): pp. 332–339.

[3] S. Peng, M. S. Qureshi, P. E. Hasler, A. Basu and F. L. Degertekin., "A Charge-Based Low-Power High-SNR Capacitive Sensing Interface Circuit". *IEEE Transactions on Circuits and Systems I: Regular Papers*, 2008. 55(7): pp. 1863–1872.

[4] J. Shah, H. Johari, A. Sharma and F. Ayazi, "CMOS ASIC for MHz silicon BAW gyroscope," *2008 IEEE International Symposium on Circuits and Systems*, Seattle, WA, 2008, pp. 2458–2461.

[5] J. A. Geen, S. J. Sherman, J. F. Chang and S. R. Lewis, "Single-chip surface micromachined integrated gyroscope with 50°/h Allan deviation". *IEEE Journal of Solid-State Circuits*, 2002. 37(12): pp. 1860–1866.

[6] M. Riverola, G. Sobreviela, F. Torres, A. Uranga and N. Barniol., "Single-Resonator Dual-Frequency BEOL-Embedded CMOS-MEMS Oscillator With Low-Power and Ultra-Compact TIA Core". *IEEE Electron Device Letters*, 2017. 38(2): pp. 273–276.

[7] C. Comi, A. Corigliano, G. Langfelder, A. Longoni, A. Tocchio and B. Simoni., "A Resonant Microaccelerometer With High Sensitivity Operating in an Oscillating Circuit". *Journal of Microelectromechanical Systems*, 2010. 19(5): pp. 1140–1152.

[8] T. Chen, J. Huang and Y. Peng *et al.*, "A 17.6-MHz 2.5V ultra-low polarization voltage MEMS oscillator using an innovative high gain-bandwidth fully differential trans-impedance voltage amplifier". in *2013 IEEE 26th International Conference on Micro Electro Mechanical Systems (MEMS)*. 2013.

[9] A. Sharma, M. F. Zaman and F. Ayazi., "A 104-dB Dynamic Range Transimpedance-Based CMOS ASIC for Tuning Fork Microgyroscopes". *IEEE Journal of Solid-State Circuits*, 2007. 42(8): pp. 1790–1802.

[10] M. Saukoski, L. Aaltonen, K. Halonen and T. Salo., "Fully integrated charge sensitive amplifier for readout of micromechanical capacitive sensors". in *2005 IEEE International Symposium on Circuits and Systems*, Kobe, 2005, Vol. 6, pp. 5377–5380.

[11] R. R. Harrison and C. Charles., "A low-power low-noise CMOS for amplifier neural recording applications". *IEEE Journal of Solid-State Circuits*, 2003. 38(6): pp. 958–965.

[12] J. Salvia, P. Lajevardi, M. Hekmat and B. Murmann, "A 56MΩ CMOS TIA for MEMS applications," *2009 IEEE Custom Integrated Circuits Conference*, Rome, 2009, pp. 199–202.

[13] B. Razavi., "A 622Mb/s 4.5pA/$\sqrt{Hz}$ CMOS transimpedance amplifier". in *2000 IEEE International Solid-State Circuits Conference. Digest of Technical Papers (Cat. No.00CH37056)*, San Francisco, CA, USA, 2000, pp. 162–163.

[14] C. T. Nguyen and R. T. Howe., "An integrated CMOS micromechanical resonator high-Q oscillator". *IEEE Journal of Solid-State Circuits*, 1999. 34(4): pp. 440–455.

[15] R. P. Leland., "Mechanical-thermal noise in MEMS gyroscopes". *IEEE Sensors Journal*, 2005. 5(3): pp. 493–500.

[16] D. B. Leeson., "A simple model of feedback oscillator noise spectrum". *Proceedings of the IEEE*, 1966. 54(2): pp. 329–330.

[17] A. A. Seshia et al., "A vacuum packaged surface micromachined resonant accelerometer". *Journal of Microelectromechanical Systems*, 2002. 11(6): pp. 784–793.

[18] T. H. Lee and A. Hajimiri., "Oscillator phase noise: a tutorial". *IEEE Journal of Solid-State Circuits*, 2000. 35(3): pp. 326–336.

[19] Y. Zhao, J. Zhao and X. Wang *et al.*, "A Sub-g Bias-Instability MEMS Oscillating Accelerometer With an Ultra-Low-Noise Read-Out Circuit in CMOS". *Solid-State Circuits, IEEE Journal of*, 2015. 50(9): pp. 2113–2126.

[20] M. Li, K. Tseng, C. Liu, C. Chen and S. Li., "An 8V 50μW 1.2MHz CMOS-MEMS oscillator". in *2016 IEEE International Frequency Control Symposium (IFCS)*, New Orleans, LA, 2016.

[21] M. Li, C. Chen, C. Liu and S. Li., "A Sub-150-BEOL-Embedded CMOS-MEMS Oscillator With a 138-dB Ultra-Low-Noise TIA". *IEEE Electron Device Letters*, 2016. 37(5): pp. 648–651.

[22] G. K. Balachandran, V. P. Petkov, T. Mayer and T. Balslink., "A 3-Axis Gyroscope for Electronic Stability Control With Continuous Self-Test". *IEEE Journal of Solid-State Circuits*, 2016. 51(1): pp. 177–186.

[23] L. He, Y. P. Xu and M. Palaniapan., "A CMOS Readout Circuit for SOI Resonant Accelerometer With 4-μg Bias Stability and 20-μg$\sqrt{Hz}$ Resolution". *Solid-State Circuits, IEEE Journal of*, 2008. 43(6): pp. 1480–1490.

[24] C. C. Enz and G. C. Temes., "Circuit techniques for reducing the effects of op-amp imperfections: autozeroing, correlated double sampling, and chopper stabilization". *Proceedings of the IEEE*, 1996. 84(11): pp. 1584–1614.

[25] M. Paavola, M. Kamarainen, J. A. M. Jarvinen, M. Saukoski, M. Laiho and K. A. I. Halonen., "A Micropower Interface ASIC for a Capacitive 3-Axis Micro-Accelerometer". *IEEE Journal of Solid-State Circuits*, 2007. 42(12): pp. 2651–2665.

[26] V. P. Petkov, G. K. Balachandran and J. Beintner., "A Fully Differential Charge-Balanced Accelerometer for Electronic Stability Control". *IEEE Journal of Solid-State Circuits*, 2014. 49(1): pp. 262–270.

[27] H. Xu, X. Liu and L. Yin., "A Closed-Loop ΣΔ Interface for a High-Q Micromechanical Capacitive Accelerometer With 200 ng/$\sqrt{Hz}$ Input Noise Density". *Solid-State Circuits, IEEE Journal of*, 2015. 50(9): pp. 2101–2112.

[28] M. Lemkin and B. E. Boser., "A three-axis micromachined accelerometer with a CMOS position-sense interface and digital offset-trim electronics". *IEEE Journal of Solid-State Circuits*, 1999. 34(4): p. 456–468.

[29] X.S. Jiang, J. I. Seeger, M. Kraft and B. E. Boser, "A monolithic surface micromachined Z-axis gyroscope with digital output", *2000 Symposium on VLSI Circuits. Digest of Technical Papers (Cat. No.00CH37103)*, Honolulu, HI, USA, 2000, pp. 16–19.

[30] A. Tocchio, A. Caspani and G. Langfelder., "Mechanical and Electronic Amplitude-Limiting Techniques in a MEMS Resonant Accelerometer". *IEEE Sensors Journal*, 2012. 12(6): pp. 1719–1725.

[31] J. Verd, A. Uranga, J. Segura and N. Barniol., "A 3V CMOS-MEMS oscillator in 0.35um CMOS technology". in *2013 Transducers & Eurosensors XXVII: The 17th International Conference on Solid-State Sensors, Actuators and Microsystems (TRANSDUCERS & EUROSENSORS XXVII)*, Barcelona, 2013, pp. 806–809.

[32] G. Sobreviela, M. Riverola, F. Torres, A. Uranga and N. Barniol., "Optimization of the Close-to-Carrier Phase Noise in a CMOS–MEMS Oscillator Using a Phase Tunable Sustaining-Amplifier". *IEEE Transactions on Ultrasonics, Ferroelectrics, and Frequency Control*, 2017. 64(5): pp. 888–897.

[33] A. Uranga, G. Sobreviela, M. Riverola, F. Torres and N. Barniol., "Design of self-sustained CMOS amplifiers for all-CMOS MEMS based oscillators". in *2016 IEEE International Conference on Electronics, Circuits and Systems (ICECS)*, Monte Carlo, 2016.

[34] A. Tocchio, A. Caspani, G. Langfelder, A. Longoni and E. Lasalandra., "A Pierce oscillator for MEMS resonant accelerometer with a novel low-power amplitude limiting technique". in *2012 IEEE International Frequency Control Symposium Proceedings*, Baltimore, MD, 2012.

[35] X. Wang, Y. P. Xu *et al.* "A 27μW MEMS silicon oscillating accelerometer with 4μg bias instability and 10μg/$\sqrt{Hz}$ noise floor". in *2018 IEEE International Symposium on Inertial Sensors and Systems (INERTIAL)*, Moltrasio, 2018.

3

MEMS Silicon Oscillating Accelerometer Readout Circuit

Xi Wang[1] and Yong Ping Xu[2]

[1]School of Information and Control Engineering, China University of Mining and Technology, Jiangsu, China
[2]Department of Electrical & Computer Engineering, National University of Singapore, Singapore
E-mail: icwangxi@cumt.edu.cn; yongping@ieee.org

This chapter describes the design of a fully differential CMOS continuous-time readout circuit for an MEMS silicon oscillating accelerometer (SOA). This chapter first introduces the concept of SOA, the existing readout circuit architectures, and the key performance parameters of the SOA. The relationship between the performance and the circuit design parameters are discussed. Based on the oscillator circuit architecture that employs automatic amplitude control (AAC), the $1/f^3$ and $1/f^2$ phase noise models are proposed and analyzed. An SOA readout circuit implemented with the proposed two-stage front-end TIA, low $1/f$ noise AAC circuit and linear VGA is then described. The dominant $1/f$ noise source that induces frequency fluctuations through amplitude-stiffness effect is found to be from the AAC circuit. The effects of $1/f$ noise are classified into additive and multiplicative components, which are suppressed by chopper stabilization and tail current source free circuit structures, respectively. The implemented SOA achieves a bias-instability of 0.4 µg, a bias-stability of 4.13 µg and a noise floor of 1.2 $\mu g/\sqrt{Hz}$ with a scale factor of 280 Hz/g and full scale of ±20 g. The chip is fabricated in 0.35-µm standard CMOS technology and consumes 4.37 mW under a 1.5 V supply.

3.1 Introduction

3.1.1 Concept of MEMS Silicon Oscillating Accelerometer (SOA)

Unlike MEMS capacitive accelerometers, in which the sensing electrode capacitance changes directly represent the input acceleration, the MEMS silicon oscillating accelerometers (SOA) is based on force-induced frequency modulation (FM) sensing principle. Figure 3.1(a) shows a simplified schematic of a MEMS resonator, which consists of a double-ended-tuning-fork-based resonant beam, and drive and sense electrodes formed by two comb structures. The input acceleration along the y-axis applies an axial load on the resonant beam and modulates its resonant frequency. By forming an MEMS oscillator with a readout circuit, the resonant frequency change induced by the input acceleration can be observed. Figure 3.1(b) shows the complete MEMS SOA system consisting of two resonators on opposite sides of the proof mass, whose resonant frequencies change differentially with the input acceleration. The final output of the SOA is the difference between the oscillation frequencies of the two MEMS oscillators formed with the two MEMS resonators, respectively. Since the output of the SOA is frequency, phase or frequency noise of the readout circuit is a prime design concern. Compared with capacitive MEMS accelerometer, the advantages of SOA can be summarized as follows:

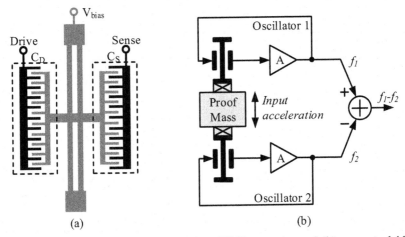

(a)   (b)

Figure 3.1 (a) Simplified schematic of the MEMS resonator and (b) conceptual block diagram of the MEMS SOA system with its readout circuit.

- Axially loading of the input force allows a large input full scale;
- Effective open-loop operation under high Q factor, resulting in a low-phase noise and hence the high bias stability;
- Time-domain signal processing allows low supply voltage without sacrificing the dynamic range of the accelerometer.

3.1.2 Readout Circuits for MEMS SOA

An MEMS oscillator-based readout circuit is usually adopted for a SOA. A typical MEMS oscillator consists of an oscillation-sustaining circuit with an amplitude-limiting or stabilization circuit. Research on the readout circuit has been mainly focused on the amplitude stabilization techniques and noise reduction.

Figure 3.2 shows four different readout circuit architectures for MEMS SOAs. In Figure 3.2(a), the amplitude of the oscillator is controlled by tuning the gain of the front-end transimpedance amplifier (TIA) [1–3]. An external reference voltage is used to set the desired amplitude. The error between the

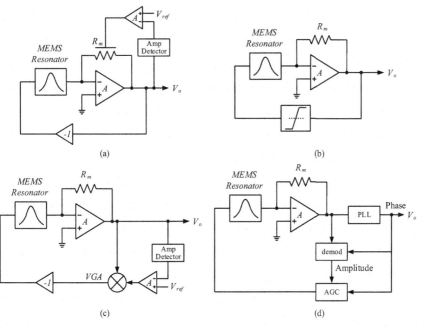

Figure 3.2 Block diagrams of SOA readout circuit architectures with (a) tuneable front-end gain, (b) comparator, (c) tuneable feedback path, and (d) PLL and AGC.

oscillator output amplitude and the external reference voltage is fed back and regulates the front-end TIA gain to stabilize the output amplitude. Although the output of the front-end (oscillator output) is stabilized, since the front-end TIA gain varies, the displacement amplitude of the MEMS resonator beam is not fixed, which could still exceed a desired amplitude and enter its nonlinear region.

In Figure 3.2(b), a comparator is inserted after the front-end amplifier in the feedback path to provide the phase for the feedback signal, while the drive amplitude is set by the output of the drive amplifier [4, 5] or a potential divider [6]. The amplitude is not closed-loop regulated in this circuit and therefore may be subject to process, supply voltage, and temperature (PVT) variations.

In Figure 3.2(c), a variable gain amplifier (VGA) is inserted in the feedback path and drives the MEMS resonator [7]. Instead of tuning the gain of the front-end amplifier, the amplitude control is achieved by adjusting the gain of the VGA. Since the gain of the front-end amplifier is fixed and its output is stabilized by an automatic amplitude control (AAC) loop, the displacement amplitude of the resonant beam is therefore fixed and well controlled. If any MEMS resonator parameter changes under environmental perturbations, resulting in a change in displacement, the AAC loop will regulate the feedback driving signal so that the displacement amplitude of the resonant beam remains unchanged. Furthermore, this circuit has separate loops for oscillator and the AAC, and the fixed gain from the front-end amplifier. It is therefore easy to optimize the circuit performance, such as noise and power.

In Figure 3.2(d) [8], the amplitude control scheme is similar to that in (c), except that the external reference is absent. An AGC, whose gain is controlled by the demodulated amplitude, regulates the feedback drive signal. A phase-locked loop (PLL) is employed to extract the output frequency information and provides the feedback drive signal with the correct phase. Due to the absence of a reference, the drive amplitude is regulated by the loop gain of the oscillator and is not well controlled. In this case, the displacement amplitude of the MEMS resonator beam is not fixed and could enter its nonlinear region, resulting in phase noise degradation. In the following section, we will focus on the AAC scheme in Figure 3.2(c).

3.1.3 Acceleration Noise Characterization

The noise at the output of the SOA dictates its performance. Three performance metrics are used to characterize the noise of the MEMS SOA, namely, bias-stability, bias-instability, and acceleration noise density. All these three metrics are measured at zero input acceleration. The bias-stability is defined as the one-sigma (σ) value of the accelerometer output data acquired within a given time duration. The bias-instability is defined as the floor of the root Allan Variance (AVAR) plot and can be regarded as the lowest instability of the bias that an accelerometer can achieve. The accelerometer noise density is obtained from the output power spectrum density (PSD) at zero input acceleration, which is usually referred to the input of the accelerometer and given as an input-referred acceleration noise density. The bias of an accelerometer represents the error near DC. In inertial navigation applications, the velocity and distance of an object are obtained by integration of its acceleration over time. Thus, the bias error will be accumulated over time and cause a significant error. For inertial navigation applications, the bias drift is desirably under μg (micro-gravity). On the other hand, the acceleration noise floor, which determines the resolution of the SOA is required to be in the order of $\mu g/\sqrt{\text{Hz}}$ or lower.

Since the readout circuit of the SOA is based on two MEMS oscillators with the input acceleration being modulated onto the oscillating frequency, the phase noise of the MEMS oscillator determines the performance of the SOA. The same methods that are used to evaluate an oscillator can be applied to characterize the SOA. Since the output of the SOA is frequency, it is more convenient to use a frequency power spectrum to evaluate its performance. The output phase noise power spectrum ($S_\phi(\Delta f)$) can be converted to the output frequency power spectrum ($S_f(\Delta f)$) as follows:

$$S_f(\Delta f) = S_\phi(\Delta f) \cdot (\Delta f)^2 \tag{3.1}$$

Further dividing the frequency noise spectrum by the scale factor of the SOA, the input-referred acceleration noise spectrum can be obtained.

Another way to characterize the accelerometer is a time-domain approach by the Allan Variance (AVAR) or root AVAR [9]. AVAR is a two-sample variance calculated from the instantaneous output of an accelerometer. A unique relationship exists between the accelerometer output PSD and AVAR [10]. The bias-instability, which is a representative of low frequency noise, is defined as the floor of the AVAR [9]. The section with slope of $-1/2$ on the AVAR plot corresponds to the white noise in the frequency noise spectrum, which is also referred to as velocity random walk (VRW).

Figure 3.3 shows the relationships among phase PSD, frequency PSD, and the AVAR, where detailed derivations can be found in [11]. It can be clearly seen that the bias-instability (AVAR floor) relates to the flicker frequency noise or the phase noise with a slope of (f^{-3}), while the phase noise with a slope of f^{-2} relates to the frequency noise floor or white frequency noise and is reflected by the $(\tau^{-1/2})$ segment on the AVAR plot.

The impacts of various noises in the oscillator circuit have been well studied for an electronic oscillator. According to the Linear Time Variant (LTV) phase noise model [12], flicker noise in the oscillation sustaining circuit causes the phase noise in (f^{-3}) segment of the phase noise spectrum (Figure 3.3(a)) or the flicker frequency noise (Figure 3.3(b)), which is the floor of AVAR plot (Figure 3.3(c)), dictates the bias-instability, while the thermal noise in the oscillator determines the phase noise in (f^{-2}) segment or the white frequency noise, and hence the resolution of the SOA. The phase noise appearing at very low frequency in the phase noise spectrum or the random walk frequency noise is caused by a slow drift of the output over

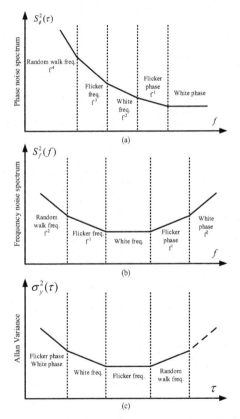

Figure 3.3   Relationships among (a) phase noise power spectrum, (b) frequency noise power spectrum, and (c) AVAR of an oscillator [11].

time, such as the drift due to slow temperature change, and shows up at large sample interval (τ) on the AVAR plot.

3.2 Readout Circuit

3.2.1 MEMS Oscillator

3.2.1.1 Front-end amplifier

The function of the front-end amplifier in the MEMS oscillator, such as those shown in Figure 3.2, is to sense and amplify the motion current from the MEMS resonator and convert it into voltage. A transimpedance amplifier (TIA) is usually employed as the front-end amplifier. It needs to provide sufficient large trans-impedance to fulfil the loop gain requirement for oscillation; meanwhile, its bandwidth should be wide enough, usually at least 10 times higher than the oscillation frequency, to avoid the excess phase shift from the TIA, which will cause the oscillation frequency to deviate from the peak of the resonance (the highest Q point). Another requirement for the TIA is the input-referred current noise. It limits the signal-to-noise ratio and sets the minimum motion current that a TIA can detect. Furthermore, even though the motion current can be detected, high input-referred noise may still impact the noise performance of the accelerometer. The input-referred noise of the TIA is usually designed to be lower than the Brownian noise of the MEMS resonator to let the Brownian noise dominate the noise at the input sense node. This can be illustrated by a simple noise model shown in Figure 3.4, where $I_{n,TIA}$ is the input-referred current noise spectrum density (A/$\sqrt{\text{Hz}}$) of the TIA, R_m is the equivalent resistance of the MEMS resonator at resonant frequency, and N_{MEMS} is the Brownian voltage noise from the MEMS resonator. Then the TIA input-referred current noise, $I_{n,TIA}$, should satisfy $I_{n,\text{TIA}} < \sqrt{4k_BT/R_m}$ so that it does not dominate the noise at the input.

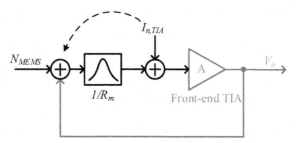

Figure 3.4 MEMS mechanical Brownian noise and the TIA input-referred noise.

3.2.1.2 Oscillation-sustaining circuit

The oscillation-sustaining circuit is essentially a positive feedback circuit that satisfies the Barkhausen Criterion and thus forms an oscillator with the MEMS resonator. Figure 3.5 shows the block diagram of the MEMS oscillator circuit. The loop gain of the oscillator is provided by a front-end TIA and a VGA in the feedback path. The output amplitude of the oscillator is extracted by an amplitude detector and fed to the input of an error amplifier. The other input of the error amplifier is connected to an external low noise voltage reference, which is used to set the amplitude of the feedback signal that drives the MEMS resonator. This in turn sets the displacement amplitude of the resonator beam and prevents it from entering the strong nonlinear region. The difference between the extracted oscillator amplitude and V_{ref} is amplified by the error amplifier whose output controls the gain of the VGA. This forms an AAC loop to regulate the loop gain of the oscillator and the feedback drive signal amplitude. Depending on the type of TIA, a phase shift of 90° may be needed after the front-end TIA to satisfy the phase condition of the oscillator.

3.2.2 Amplitude Control

3.2.2.1 Amplitude-stiffness effect

In an LTV phase noise model for electronic oscillators, the amplitude-induced phase noise is ignored with an assumption that the amplitude is well regulated by the amplitude control circuit. For MEMS-based oscillators, however, due

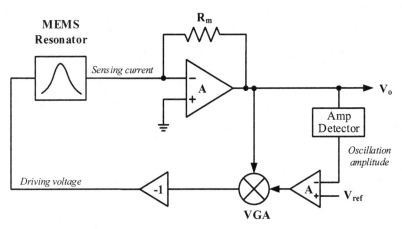

Figure 3.5   Simplified block diagram of the MEMS oscillator circuit.

to the stiffness nonlinearity of the MEMS resonator beam, the nonlinear stiffness term is a function of the MEMS resonator beam displacement amplitude. This implies that the resonant frequency of the MEMS resonator can be modulated by the amplitude noise of the oscillator. Thus, in an MEMS-based oscillator, the amplitude noise cannot be neglected. As described in [13], the nonlinear stiffness of the MEMS resonator is modelled by a cubic displacement term in the motion equation, and the resonant frequency dependence on the displacement amplitude can be expressed as

$$f = f_0 \left(1 + \frac{3}{8} \frac{k_3}{k_1} X_0^2 \right) \tag{3.2}$$

where k_1 and k_3 are the linear and cubic elastic stiffness coefficients of the resonant beam, respectively, and X_0 is the oscillation amplitude. Hence, a small deviation in the oscillation amplitude will affect the resonant frequency due to the nonlinear stiffness, which is given by

$$\Delta f = f_0 \cdot \frac{3}{4} \frac{k_3}{k_1} X_0 \Delta x \tag{3.3}$$

This is often referred to as the amplitude-stiffness (A-S) effect [14]. Equation (3.2) can also be expressed in terms of power spectrum, as shown in Equation (3.3), where $S_A(\Delta f)$ is the amplitude noise PSD and $S_f^2(\Delta f)$ is the frequency PSD,

$$S_f^2(\Delta f) = \frac{9}{16} \left(\frac{k_3}{k_1} \right)^2 f_0^2 X_0^2 \cdot S_A^2(\Delta f) \tag{3.4}$$

We can observe that the flicker noise component in the amplitude can be directly converted to flicker frequency noise, which consequently deteriorates the bias of the accelerometer.

3.2.2.2 Noise model of the AAC loop

The completed block diagram of the MEMS oscillator in the SOA with all noise sources is shown in Figure 3.6. The driving signal, V_{VGA}, is multiplied with V_{BIAS} to generate electro-static force and drives the MEMS resonator. N_{BIAS} and N_{REF} are the input voltage noise from V_{BIAS} and external V_{REF}, respectively. N_{MEMS} is the equivalent mechanical thermal voltage noise of the MEMS transducer ($v4k_BTR_m$). N_{TIA} is the input-referred current noise of the front-end amplifier. N_{AAC} is the lumped voltage noise of the AAC circuit, including the error amplifier and the amplitude detector, and N_{VGA} is the output-referred voltage noise of the VGA.

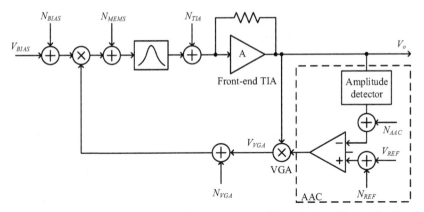

Figure 3.6 Complete block diagram of the MEMS oscillator in the SOA with all noise sources [23]. (Courtesy of IEEE).

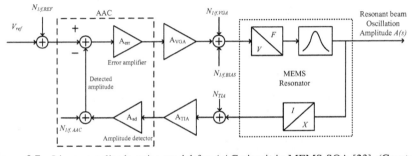

Figure 3.7 Linear amplitude noise model for AAC circuit in MEMS SOA [23]. (Courtesy of IEEE).

Based on Figure 3.6, a linear noise model of the displacement amplitude can be derived, as shown in Figure 3.7. The AAC block works in the low frequency region, while the oscillation loop works at the resonant frequency. The input is the reference voltage (V_{REF}), and the output is the displacement amplitude of the resonant beam, denoted by A(s). $K_{V/F}$ is the conversion gain from voltage to electro-static force, and $K_{X/I}$ is the conversion gain from the displacement of the resonant beam to the motion current. All the gains can be treated as constant coefficients at their working frequencies. Note that this model is for amplitude only, that is, in the absence of all noises, the output should be a constant displacement amplitude in meter. Since the amplitude path works in a low frequency region, flicker noise becomes the dominant noise source. Thus, for simplicity, most noises in Figure 3.6 are replaced by flicker noise only, except the noise from the front-end, N_{TIA}, which contains both flicker and thermal noises. The flicker noise in the front-end amplifier is phase modulated (PM) to the carrier frequency, which means that it does not affect the amplitude of the oscillation signal. Besides, the PM flicker noise component is not significant due to symmetrical design and high Q of the resonator [24]. Thus, it is neglected in further discussion. The remaining flicker noise sources are the noise in V_{REF} ($N_{1/f,REF}$), the lumped noise in AAC circuit ($N_{1/f,AAC}$), the output noise of VGA ($N_{1/f,VGA}$) and the noise in MEMS bias voltage ($N_{1/f,BIAS}$). $N_{1/f,VGA}$ represents the flicker noise component that affects the gain of the VGA and consequently the amplitude noise. Note that in Figure 3.6, the electrostatic driving force is generated by multiplying the voltage difference between V_{BIAS} and V_{VGA}, which means that the $N_{1/f,BIAS}$ is effectively modulated to the oscillation frequency.

Under an assumption that the loop gain is much greater than one, which can always be realized with large error amplifier gain, Hence, the closed-loop transfer functions from input, V_{ref}, to the output can be approximated by the reciprocal of the feedback gain, that is, $(A_{ad}A_{TIA}K_{X/I})^{-1}$. Thus, the noise transfer functions can be derived and given below:

$$\frac{A(s)}{N_{1/f,REF}(s)} \approx \frac{1}{A_{ad}A_{TIA}K_{X/I}}$$

$$\frac{A(s)}{N_{1/f,AAC}(s)} \approx \frac{1}{A_{ad}A_{TIA}K_{X/I}}$$

$$\frac{A(s)}{N_{1/f,VGA}(s)} = \frac{A(s)}{N_{1/f,BIAS}(s)} \approx \frac{1}{A_{ad}A_{TIA}K_{X/I}} \cdot \frac{1}{A_{err}A_{VGA}} \qquad (3.5)$$

It can be seen that $N_{1/f,REF}$ and $N_{1/f,AAC}$ are both located at the input node of the loop and is directly appear on the amplitude (at the output) after amplified by the closed-loop gain. The VGA and bias noises, $N_{1/f,VGA}$ and $N_{1/f,BIAS}$, are attenuated by the gain of the error amplifier and VGA. Thus, to reduce the flicker noise in the displacement amplitude, the flicker noises in the external reference and AAC circuit, $N_{1/f,REF}$ and $N_{1/f,AAC}$, must be minimized. High error amplifier gain also helps reduce the noise contributions from $N_{1/f,BIAS}$ and $N_{1/f,VGA}$. Note that the VGA gain cannot be arbitrarily increased, since it also determines the loop gain of the MEMS oscillator.

3.2.3 Phase Noise of the MEMS Oscillator

Phase noise in an electronic oscillator has been well studied, and the LTV phase noise model in [12] gives accurate prediction of the phase noise and its origins. As discussed in Section 3.1.3, the flicker noise in the oscillator dominates the bias-instability, while the thermal noise determines the acceleration resolution. However, if the oscillator has a balanced architecture, the flicker noise-induced phase noise in $1/f^3$ region can be ignored [12]. Nevertheless, from the discussion in Section 3.2.2.1, the oscillator amplitude noise can be directly translated to the frequency noise through the A-S effect. Thus, the flicker noise in the amplitude becomes a dominant source for the $1/f^3$ phase noise, or the $1/f$ frequency noise, and therefore dominates the bias-instability of the accelerometer.

The following discussion will be focused on the phase noise in the $1/f^2$ region, which determines the resolution of the SOA. For simplicity, the LTI phase noise model [15, 16] is used for the analysis here.

A simplified block diagram of an MEMS oscillator is shown in Figure 3.8, where N_{MEMS} is the equivalent thermal voltage noise from the MEMS resonator, N_{TIA} is the output-referred voltage noise of the front-end amplifier, and N_1 is the total output voltage noise of AAC circuit and VGA. $H(j\omega)$ is the equivalent transfer function of the MEMS resonator, which includes the conversion gains at its interfaces. The noise transfer functions can be easily obtained:

$$\frac{Y(j\omega)}{N_{TIA}(j\omega)} = \frac{1}{1 - A_1 A_2 H(j\omega)}$$

$$\frac{Y(j\omega)}{N_{MEMS}(j\omega)} = \frac{Y(j\omega)}{N_1(j\omega)} = \frac{A_1 H(j\omega)}{1 - A_1 A_2 H(j\omega)} \quad (3.6)$$

The noise, N_1, injects at the same node as the mechanical Brownian noise, and therefore has the same noise transfer function. To gain more insights, the derivation from the transfer functions to side-band noise transfer function can be found in [15] and also explained below. At a small frequency offset ($\Delta\omega$) from the oscillation frequency, the transfer function of the resonator can be

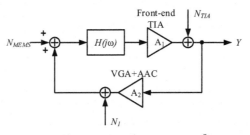

Figure 3.8   Simplified oscillator model of the SOA for $1/f^2$ phase noise calculation.

approximated to $H(j\omega_0) + \Delta\omega \cdot dH/d\omega$ and $\Delta\omega \cdot dH/d\omega \ll 1$. Thus, the transfer functions of noise power can be derived from Equation (3.6):

$$\frac{Y(j\omega + \Delta\omega)}{N_{TIA}(j\omega + \Delta\omega)} = -\frac{1}{A_1 A_2 \Delta\omega \frac{dH}{d\omega}}$$

$$\frac{Y(j\omega + \Delta\omega)}{N_{MEMS}(j\omega + \Delta\omega)} = \frac{Y(j\omega + \Delta\omega)}{N_1(j\omega + \Delta\omega)} = -\frac{1/A_2}{A_1 A_2 \Delta\omega \frac{dH}{d\omega}} \qquad (3.7)$$

Let $H(j\omega) = A(\omega)e^{j\Phi(\omega)}$, hence dH/d$\omega$ can be expressed as:

$$\frac{dH}{d\omega} = \left(\frac{dA}{d\omega} + jA\frac{d\Phi}{d\omega}\right)e^{j\Phi} \qquad (3.8)$$

By defining Q as

$$Q = \frac{\omega_0}{2}\sqrt{\left(\frac{dA}{d\omega}\right)^2 + \left(\frac{d\Phi}{d\omega}\right)^2} \qquad (3.9)$$

we can obtain Equation (3.10) showing the transfer function from noise sources in the loop to side-band noise power [15],

$$\left|\frac{Y(j\omega)}{N_{TIA}(j\omega)}\right|^2 \approx \left|\frac{1}{A_1 A_2 \Delta\omega \frac{dH}{d\omega}}\right|^2 \approx \left(\frac{\omega_0}{2Q}\right)^2 \frac{1}{\Delta\omega^2}$$

$$\left|\frac{Y(j\omega)}{N_{MEMS}(j\omega)}\right|^2 = \left|\frac{Y(j\omega)}{N_1(j\omega)}\right|^2 \approx \left|\frac{1/A_2}{A_1 A_2 \Delta\omega \frac{dH}{d\omega}}\right|^2 \approx \left(\frac{1}{A_2}\frac{\omega_0}{2Q}\right)^2 \frac{1}{\Delta\omega^2}$$

$$(3.10)$$

which presents a slope of f^{-2}. On the other hand, at large frequency offset, $H(j\omega)$ in Equation (3.6) gets smaller and making the noise from N_{MEMS} and N_1 smaller than the system noise floor, so the only remaining noise is N_{TIA}. The total noise power can be expressed as

$$Y^2(s) = \left(\frac{\omega_0}{2Q}\frac{1}{\Delta\omega}\right)^2 \left(\frac{N^2_{MEMS}(s) + N^2_1(s)}{A_2^2} + N^2_{TIA}(S)\right) + N^2_{TIA}(s)$$

$$(3.11)$$

The first term represents the $1/f^2$ sideband noise and the second term represents the white noise floor. It can be seen that the noise of the front-end amplifier and the feedback circuits also contribute to the $1/f^2$ sideband noise region. In order to let the noise of the MEMS resonator dominate the performance, the N_1 and input-referred N_{TIA} must be made smaller than N_{MEMS}.

The noise in (3.11) is the noise at the output of the oscillator and can be referred to the input of the TIA and related to the motion current, I_{sense}, from the MEMS resonator. Assuming that the $1/f^2$ phase noise is dominated by N_{MEMS} (i.e. $N_1 \ll N_{MEMS}$) and the phase noise is half of the sideband noise, the phase noise spectrum can be obtained at the input sensing node by dividing the total input-referred noise by the full-scale motion current power, I^2_{sense}

$$S^2_{\phi}(\Delta\omega) = \frac{1}{2}\left[\left(\frac{\omega_0}{2Q}\frac{1}{\Delta\omega}\frac{\sqrt{4k_BT/R_m}}{I_{sense}}\right) + \frac{I^2_{n,TIA}}{I^2_{sense}}\right] \qquad (3.12)$$

where $I_{n,TIA}$ is the input-referred current noise of the TIA and I_{sense} is the root-mean-square (RMS) amplitude of the motion current from the resonator. As described in Section 3.1.3, the $1/f^2$ phase noise spectrum determines the noise floor in the frequency noise spectrum. This equation implies that the acceleration noise floor can be improved with high quality factor, large proof mass and scale factor.

3.3 Circuit Implementation

3.3.1 Overall Readout Circuit at System Level

This section describes the implementation of the MEMS SOA readout circuit. The complete system block diagram of a single readout channel is shown in Figure 3.9. The MEMS transducer consists of two resonators on opposite sides of the proof mass, which enables differential sensing and common-mode noise rejection. The readout circuit consists of two readout channels and each of them forms an oscillator with the MEMS resonators. The front-end amplifier is a continuous-time two-stage trans-impedance amplifier, comprising an integrator followed by a differentiator. The AAC circuit consists of a buffer, an amplitude detector, an error amplifier and a loop filter. The VGA, whose gain is controlled by the AAC circuit output, regulates the oscillator loop gain and drives the MEMS resonator, completing the oscillator loop.

3.3.2 Front-end Amplifier

As discussed in Section 3.2, the three key requirements for the front-end amplifier are high gain, low input-referred noise, and negligible excess phase

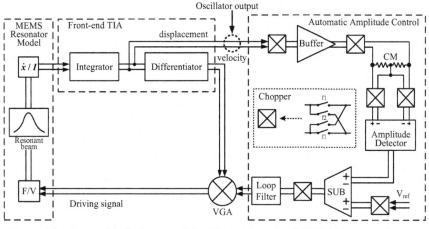

Figure 3.9 System block diagram of the SOA readout circuit [21]. (Courtesy of IEEE).

shift. The front-end amplifier mainly contributes to $1/f^2$ phase noise or white frequency noise that dictates the SOA resolution. Since the double-ended tuning fork (DETF) MEMS resonator is fully differential, resulting in a symmetrical oscillator output, the flicker noise in the front-end amplifier can be ignored, whereas the $1/f^3$ phase noise is dominated by the flicker noise from the AAC circuit.

Conventional front-end amplifiers adopt either a single-stage TIA or charge-sensing amplifier (CSA), as shown in Figure 3.10. The transimpedance of the TIA and CSA are R_f and $1/\omega_0 C_f$, respectively, where ω_0 is the oscillation frequency. C_c in Figure 3.10(a) is for frequency compensation, R_b in Figure 3.10(b) provides a DC feedback, and C_p is a parasitic capacitance at the input of each amplifier. The TIA needs to compromise between high gain, bandwidth and stability due to the parasitic capacitance at the input [22], while the CSA has a 90° "excess" phase shift due to its integrator nature. A two-stage bandpass TIA in Figure 3.11 [17] relaxes the above-mentioned trade-off and provides a correct phase for the oscillator. The first stage, similar to the CSA, senses the motion current from the MEMS resonator and integrates it on the feedback capacitor, C_1. Its voltage output is proportional to the displacement of the resonant beam. The input common-mode is set by its output through a DC feedback pseudo-resistor, R_b. The cut-off frequency of the integrator is at several Hz to ensure that it is an ideal integrator at oscillation frequency. The second stage performs differentiation to provide 90° leading phase shift so that its output is proportional to the velocity of the resonant beam. Combining the two stages, a bandpass TIA is

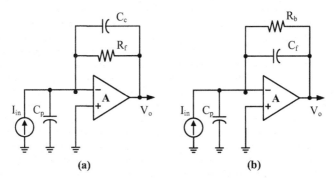

Figure 3.10   Schematic of conventional (a) TIA and (b) CSA.

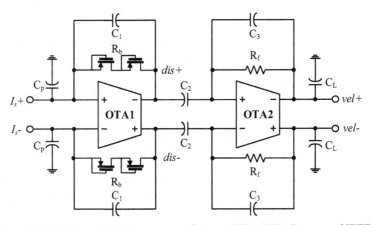

Figure 3.11   Schematic of the two-stage front-end TIA [17]. (Courtesy of IEEE).

formed, which provides a pass-band gain of (R_fC_2/C_1) and near-zero phase shift at oscillation frequency. Furthermore, the TIA provides two outputs that are proportional to the displacement and velocity of the resonator beam, respectively. The velocity output of the TIA can be directly fed back to drive the MEMS resonator, whereas the displacement output is used for AAC. This two-stage TIA has the following features:

- high gain and bandwidth trade-off is released;
- the dominant noise is from R_f and it is attenuated by C_2/C_1 when referred to input;
- the unwanted pole caused by input parasitic capacitance, C_p, is pushed to near the zero frequency and hence improves the stability; and
- both displacement and velocity signals are available.

More details of the design considerations and measurement results can be found in Chapter 2 of this book.

3.3.3 AAC Circuits

The block diagram of the AAC circuit is also shown in Figure 3.1, which consists of a buffer to reduce the loading to the front-end amplifier, an amplitude detector, an error amplifier to amplify the difference between the extracted oscillator amplitude and the external reference, and a loop filter to remove the high frequency contents. The output from the loop filter regulates the gain of the VGA whose output drives the MEMS resonator.

As discussed in Subsection 3.2.2.2, due to the A-S effect, the flicker amplitude noise is up-converted to the oscillation frequency and deteriorate the bias-instability of the SOA. Thus, the design of the AAC circuit should focus on flicker noise reduction.

Chopper stabilization is an effective technique to reduce the flicker noise [18]. However, depending on the circuit, the flicker noise can be linearly added to the input signal. It can also modulate the gain of the signal path. In the latter case, the flicker noise cannot be suppressed by the chopper stabilization technique. Taking a common source amplifier in Figure 3.4 as an example, the flicker noise of the input transistor, M_N, is linearly added to the input by superposition. On the contrary, the flicker noise in current source transistor, M_P, modulates the bias current of M_N and hence the gain of the amplifier, as given below,

$$Gain \propto \sqrt{2\mu C_{ox}\frac{W}{L}(I_B + I_n)} \cdot \frac{1}{\lambda(I_B + I_n)} \approx g_{m,0}r_{o,0}\left(1 - \frac{I_n}{2I_B}\right)$$

$$g_{m,0} = \sqrt{2\mu C_{ox}\frac{W}{L}I_B}$$

$$r_{o,0} = \frac{1}{\lambda I_B} \qquad (3.13)$$

where $g_{m,0}$ and $r_{o,0}$ are the ideal noiseless trans-conductance and output impedance, respectively, I_B is the bias current for M_N and I_n is the flicker noise-induced current noise in M_P ($I_n \ll I_B$). This equation implies that the flicker noise in the bias current influences the gain of the amplifier. Let's consider the flicker noise from M_P only and assume that transistor, M_N, has no flicker noise, and the input signal (V_{in}) is at DC. After the first chopper, V_{in} is modulated to chopper frequency, f_{ch}, before applied to the input of the amplifier (i.e., V_A). V_A is then multiplied by the amplifier gain, which is contaminated by the flicker noise from M_P ($N_{P,1/f}$), and appears in the output at the chopper frequency. After the second chopper, the output signal corrupted by $N_{P,1/f}$ is demodulated back to DC, as shown in Figure 3.12(a). This shows that the flicker noise in the bias current cannot be removed by the chopper. This is also evident in a simulation of the chopper stabilized common-source amplifier in (a) where V_{in} is connected to a DC signal, as shown in Figure 3.12(b). The solid line represents the noise power spectrum of V_O calculated by taking FFT on the time-domain output and the dashed-line represents the side-band noise power spectrum of V_B by

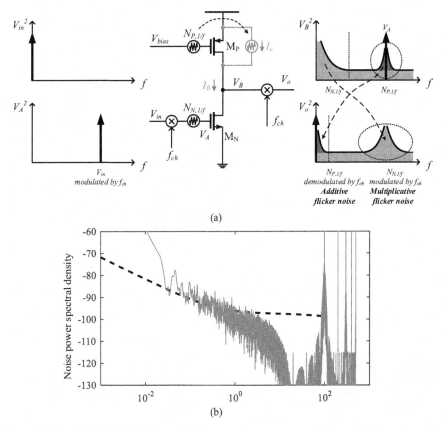

Figure 3.12   (a) Effect of flicker noise in a common source amplifier with chopper stabilization and (b) simulated noise PSD of V_O (solid-line) and side-band noise PSD of V_B (dashed-line) [23]. (Courtesy of IEEE).

PSS and PNOISE simulations. The agreement of these two spectra indicates that a part of the flicker noise is modulated to chopper frequency and that it cannot be attenuated by chopper technology. From the above analysis, the flicker noises in M_N and M_P impact differently on the amplifier performance. To distinguish these two flicker noises, they are referred to as "additive" ($N_{N,1/f}$) and "multiplicative" ($N_{P,1/f}$) flicker noise, respectively. The former is linearly added to the input signal, while the latter multiplies the input signal through gain modulation. Note that the multiplicative flicker noise cannot be attenuated by chopper stabilization.

3.3.4 Amplitude Detector

The amplitude detector extracts the amplitude of the oscillator output. Since the oscillation amplitude is usually small (around 200 mV), especially with a low supply voltage, conventional rectification circuit using diodes cannot be employed due to its high turn-on voltage. Here an alternating voltage follower (AVF) [19] shown in Figure 3.13(a) is adopted. It consists of two common drain amplifiers with a shared load. The output of the AVF always follows the input that has lower voltage (gate voltage). As discussed earlier, the flicker noise of the active load, M_{P0}, which also provides the bias current, contributes multiplicative flicker noise; while the flicker noise of the common drain amplifiers contributes additive flicker noise.

Due to the source degeneration feedback in the AVF, the gain is approximately $g_m r_o/(1+g_m r_o)$, which implies that the "noise" in gain due to the multiplicand flicker noise is suppressed for $g_m r_o \gg 1$. On the other hand, since AVF is a voltage follower, the additive flicker noises of MP_1 and

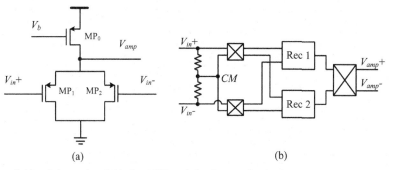

(a)   (b)

Figure 3.13 Schematic of (a) the AVF and (b) the amplitude detector using dual rectifiers [21]. (Courtesy of IEEE Press).

MP$_2$ will directly appear at the output without any attenuation. Furthermore, the output always follows the higher signal at the two input terminals, so chopping does not have any effect. The work in [7] proposes an alternative solution, in which dual rectifiers are employed, as shown in Figure 3.13(b). The common mode voltage of the differential inputs is extracted by the two resistors. The input signal and the extracted common-mode level are connected to the choppers and the two rectifiers perform rectification while providing reference to each other alternatively during the chopper operation. Therefore, the outputs of the two rectifiers are at chopping frequency and can be demodulated back to DC at the second chopper.

3.3.5 Buffer

A buffer is inserted after the front-end amplifier to isolate the resistive loading from the AAC circuit. Since the following amplitude detection circuit extracts only amplitude information which is at DC, the flicker noises from the front-end amplifier and the buffer should be minimized. Another role of the buffer is to provide a DC-level shift so that the amplitude detector can function properly. The AVF is essentially a source follower that shifts its input DC voltage up by a V_{gs}. This will limit the output signal swing. Native transistors reduce the level shift by making V_{th} approaching zero. However, they may not be available in some processes and their mismatch is usually higher than normal transistors. Thus, a hybrid structure that combines buffer and level shift is employed here.

The schematic of the hybrid buffer is shown in Figure 3.14. The buffer stage is an AC-coupled amplifier. The differential outputs of the OTA are

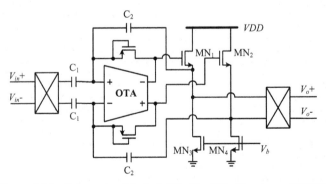

Figure 3.14   Schematic of the hybrid buffer [21]. (Courtesy of IEEE).

connected to two source followers for level shift. The feedback signal is taken from the output of the source follower, M_{N1} and M_{N2}, instead of the outputs of the OTA, to form a full closed-loop buffer. The advantages of the hybrid buffer are summarized as follows:

- The source follower is placed inside of feedback loop; thus the overall gain of the buffer is not affected by the gain of the source follower. This eliminates the gain loss of source follower if it were placed outside of the feedback loop as a standalone level shifter. The overall gain is determined by C_1/C_2;
- The effect of source follower gain variation is suppressed by the feedback, and the overall gain linearity is improved;
- The source degeneration feedback in the source follower attenuates the multiplicative flicker noise from M_{N3-4}. Moreover, it is further attenuated by the main feedback loop.

3.3.6 Error Amplifier

The input of the error amplifier is the difference between the detected amplitude signal and the external V_{REF}. There are two design considerations for the error amplifier. First, according to the amplitude control loop noise model in Figure 3.7 (in Subsection 3.2.2.2), the error amplifier provides the forward gain of the amplitude control loop and attenuate the noises from the VGA and the polarization bias voltage. Thus, in order to minimize the noise, the gain of the error amplifier should be maximized. Secondly, the flicker noise of the error amplifier is at the input node of the loop and thus will appear in the output amplitude unattenuated, and therefore should be minimized.

Figure 3.15 shows an error amplifier in [7]. It has an open-loop configuration. V_{REF} and oscillation amplitude, V_{amp}, are both converted to current by two transconductance amplifiers, G_M, and subtracted in current-domain. A chopper is applied to attenuate the flicker noise from the two G_M cells. The loop filter serves as a low pass filter and a PI controller for the AAC loop. The overall gain of the error amplifier, $|G_M(s)L(s)|$, is large at low frequency. Hence, the low-frequency noises of the VGA and the MEMS bias voltage are greatly attenuated, meeting the first design requirement.

The G_M cell used in [7] is shown in Figure 3.16(a) and its transconductance is given by

$$G_M = \frac{1}{R_s} \frac{g_m R_s}{1 + g_m R_s} \qquad (3.14)$$

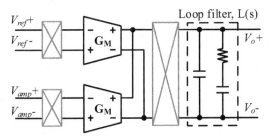

Figure 3.15   Schematic of the error amplifier with chopper stabilization in [7].

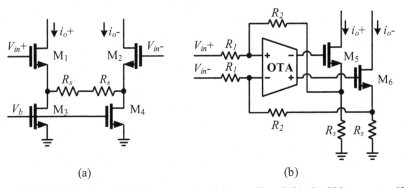

(a)   (b)

Figure 3.16   Schematic of (a) the conventional G_M cell and (b) the V-I converter [21]. (Courtesy of IEEE).

where assuming R_S is much smaller than the output resistances of M_3 and M_4. To reduce the impact of the multiplicative flicker noise from M_3 and M_4, $g_m R_S$ should be much greater than one. However, this is not practical because 1) g_m is limited by power consumption and 2) the increase of R_S reduces the transconductance of G_M. As a result, the attenuation of multiplicative flicker noise by source degeneration feedback is limited.

The role of the G_M cell is to convert the voltage to current. Alternatively, a voltage-to-current (V-I) converter can be used. Figure 3.16(b) shows a closed-loop V-I converter. The transistors that provide bias currents in Figure 3.16(a), which is the source of the multiplicative flicker noise, is replaced by resistors. In order to reduce the dependence of the V-I converter gain on $g_m R_S$, an OTA is inserted before the trans-conductance amplifier to increase the overall open-loop gain. Meanwhile the current-to-current feedback topology is employed to form a closed-loop structure. The overall trans-conductance is determined by $R_2/(R_1 R_S)$, increased by a ratio of R_2/R_1. The added OTA only contributes additive flicker noise and can be attenuated by the chopper stabilization.

The complete error amplifier with the loop filter is shown in Figure 3.17. The oscillation amplitude signal (V_{amp}) and the reference signal (V_{ref}) are converted to current by two V-I converters and subtracted. The error current signal is then converted back to the voltage domain by a type-II loop filter, which provides a large transimpedance gain at low frequency and attenuates the noise contribution from the subsequent blocks. Active loads, M_{P1}, and M_{P2} with common-mode feedback (CMFB) provides the DC path for the output branch and sets the output common-mode voltage.

The comparison between the two error amplifiers is performed by simulations of two complete AAC circuits while maintaining the other blocks the same. The output power spectrums shown in Figure 3.18 prove that the AAC using V-I converter has much lower flicker noise.

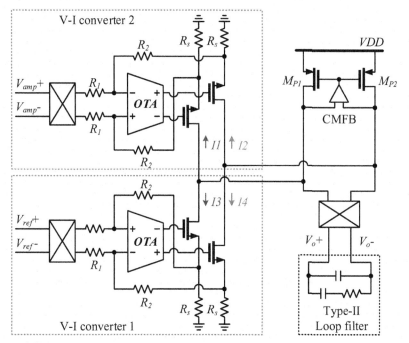

Figure 3.17 Complete schematic of the error amplifier (V-I converter) with the loop filter [21] (Courtesy of IEEE).

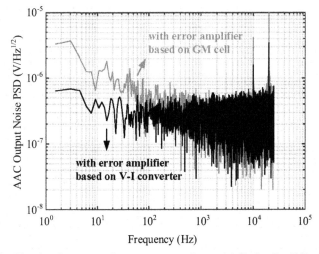

Figure 3.18 Simulated output noise spectrums of two AAC circuits [23]. (Courtesy of IEEE).

3.3.7 Variable Gain Amplifier (VGA)

The function of the VGA is to adjust the loop gain of the oscillator, based on the extracted amplitude information from the AAC circuit, so that the oscillator loop gain can be satisfied. Figure 3.19 shows the schematic of a linear VGA. The differential output of the AAC circuit, V_{c+} and V_{c-}, are connected to the gates of the foot transistors M_{N1-4}. These four transistors operate in linear regions and function as voltage controlled resistors, denoted R_S. The output of the front-end amplifier, vel_+ and vel_-, is connected to the gates of M_{N5-8}, which work in saturation regions. Thus, the gain of a single branch is approximately proportional to R_L/R_S, and the complete gain of the VGA is given by

$$R_{s+} = \frac{1}{\mu C_{ox}(V_{C+} - V_{TH})}, R_{s-} = \frac{1}{\mu C_{ox}(V_{C-} - V_{TH})}$$

$$V_o = vel_+ \frac{R_L}{R_{s+}} vel_- \frac{R_L}{R_{s-}} - vel_+ \frac{R_L}{R_s-} - vel_- \frac{R_L}{R_{s+}}$$

$$= \mu C_{ox} R_L (vel_+ - vel_-)(V_{C+} - V_{C-}) \tag{3.15}$$

where R_{s+} is the linear region resistance of M_{N1} and M_{N3}, and R_{s-} is that of M_{N2} and M_{N4}. It can be seen that the gain of the linear VGA is linearly proportional to V_C, which has a better linearity compared to Gilbert Cell [20] in the case when one of the inputs is at DC.

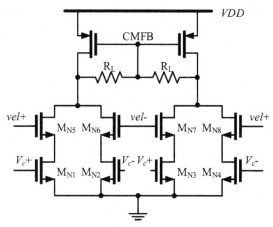

Figure 3.19 Schematic of the linear VGA [17]. (Courtesy of IEEE).

3.4 Performance

The readout circuit is implemented in a standard 0.35 um CMOS process and the chip micrograph is shown in Figure 3.20(a). The active area of the chip is 2.4 mm^2, and the area including I/O pad is 6 mm^2. The microphotograph of the MEMS transducer is shown in Figure 3.20(b). The MEMS

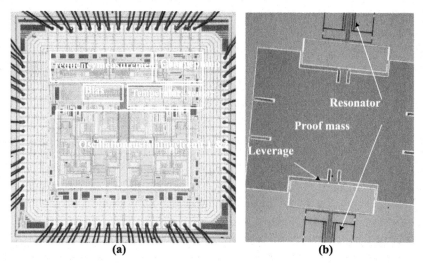

Figure 3.20 Chip micrograph of (a) the CMOS readout circuit and (b) the MEMS transducer [23]. (Courtesy of IEEE).

Table 3.1   Performance Summary of the Front-End TIA (Simulation & Measurement)

TIA Parameters	Simulation	Measurement
Transimpedance	45 MΩ	44.5 MΩ
Input-referred current noise @21 kHz	4.5 fA/$\sqrt{Hz}$	6.7 fA/$\sqrt{Hz}$
Low cut-off frequency	1 Hz	0.5 Hz
High cut-off frequency	450 kHz	350 kHz

transducer shares the same design with [17]. More details of the MEMS transducer design can be found in Chapter 1 of this book. The nominal resonant frequency, quality factor, and scale factor of the MEMS transducer are measured to be approximately 25 kHz, 30, 000, and 140 Hz/g (single resonator), respectively. The two chips are mounted on a PCB and connected via bonding wires. Drive and sense I/O pads are separated and shielded by the ground to prevent feed-through.

The performance of the front-end TIA is summarized in Table 3.1. The trans-impedance gain at the oscillation frequency is 44.5 MΩ and the passband is from 0.5 Hz to 350 kHz. The large bandwidth is to minimize the phase shift of the oscillation signal, which is measured to be less than 2.5°. The output voltage noise density is 300 nV/$\sqrt{Hz}$, corresponding to an input-referred current noise density of 6.7 fA/$\sqrt{Hz}$. As a comparison, the equivalent mechanical thermal noise current of the MEMS transducer is 90 fA/$\sqrt{Hz}$.

To measure the noise performance of the complete SOA, the device is placed on a levelled ground to prevent gravity impacts. Both readout channels are measured at the same time. Linear Least Mean Square (LMS) fitting method is applied to compensate the mismatch between the two channels in both acceleration noise floor and bias-instability measurements under influence of environmental perturbations, such as temperature drifts. Hence, the drift due to temperature effect and biasing voltage sources can be compensated to the first order.

To measure the acceleration noise floor, an FFT is performed on the measured frequency data. The acceleration noise spectrum is then divided by the scale factor of the SOA. The measured input-referred acceleration noise is shown in Figure 3.21. The measurement period is more than 1,000 seconds so that the bin width in the spectrum can be as low as 1 mHz. The acceleration noise floor measured at the oscillator output (output of front-end amplifier) is 1.2 μg/$\sqrt{Hz}$.

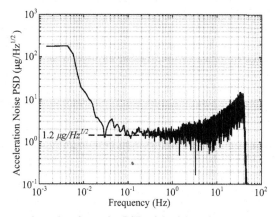

Figure 3.21 Measured acceleration noise PSD of the SOA after the first-order compensation [21]. (Courtesy of IEEE).

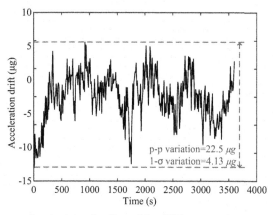

Figure 3.22 Measured output acceleration of the SOA at room temperature after the first-order compensation.

In long-term stability measurements, the total size of the measure data is very large. To reduce the computation complexity, the original output data rate (380 Hz) is decimated to 0.1 Hz, as shown in Figure 3.22. The bias-stability is obtained to be 4.13 μg (1 σ deviation). The bias-instability is obtained by calculating the AVAR on the measured data (decimated to 3.8 Hz) and is plotted in Figure 3.23. The bias-instability or the AVAR floor is 0.4 μg.

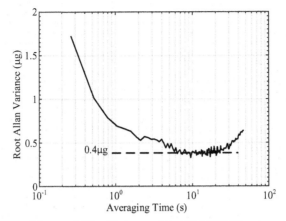

Figure 3.23   Measured Allan Variance of the SOA [21] (Courtesy of IEEE).

3.5 Conclusion

In this chapter, we have introduced the concept of the MEMS SOA, its readout circuit architecture, and key performance metrics. The impact of amplitude-stiffness effect on MEMS oscillator phase noise is discussed and the flicker noise in the amplitude control circuit is identified as a major noise source that degrades the bias-instability of the SOA. A linear noise model is proposed to analyse the contributions of various noise sources in the AAC circuit. A closed-loop error amplifier is employed to reduce both additive and multiplicative flicker noises. Finally, a complete MEMS SOA has been demonstrated in [23] and [21]. The key performance metrics are summarized in Table 3.2.

Table 3.2   Key Performance Summary of the Complete SOA

SOA Parameters	Measurement Results
Full-scale range	± 20 g
Acceleration noise density floor	$1.2 \ \mu g/\sqrt{Hz}$
Bias-instability	$0.4 \ \mu g$
Bias-stability	$4.13 \ \mu g$
Power supply	1.5 V
Power consumption	4.37 mW

References

[1] A. Roessig, et al., "Surface-micromachined resonant accelerometer," Solid State Sensors and Actuators, 1997. TRANSDUCERS '97 Chicago, 1997 International Conference on, Chicago, IL, 1997, vol. 2, pp. 859–862.

[2] A. Tocchio, et al., "A pierce oscillator for MEMS resonant accelerometer with a novel low-power amplitude limiting technique," Frequency Control Symposium (FCS), 2012 IEEE International, May. 2012, pp. 1–6.

[3] Seungbae Lee, "Influence of automatic level control on micromechanical resonator oscillator phase noise," Proc. 2003 IEEE Intl. Frequency Control Symposium.

[4] C. Comi, et al., "A Resonant Microaccelerometer With High Sensitivity Operating in an Oscillating Circuit," in Journal of Microelectromechanical Systems, vol. 19, no. 5, Oct. 2010, pp. 1140–1152.

[5] H. C. Kim, et al., "Inertial-grade out of-plane and in-plane differential resonant silicon accelerometers (DRXLs)," in the 13th International Conference on Solid-State Sensors, Actuators and Microsystems, 2005. Digest of Technical Papers. SENSORS '05, vol. 1, June 2005, pp.172–175.

[6] X. Zou and A. A. Seshia, "A high-resolution resonant MEMS accelerometer," 2015 Transducers - 2015 18th International Conference on Solid-State Sensors, Actuators and Microsystems (TRANSDUCERS), Anchorage, AK, 2015, pp. 1247–1250.

[7] L. He, Y. P. Xu, M. Palaniapan, "A CMOS readout circuit for SOI resonant accelerometer with 4μg bias stability and $20\mu\ g/\sqrt{Hz}$ resolution," IEEE J. Solid-State Circuits, vol. 43, no. 6, June 2008, pp. 1480–1490.

[8] S. A. Zotov, et al., "High Quality Factor Resonant MEMS Accelerometer with Continuous Thermal Compensation," in IEEE Sensors Journal, vol. 15, no. 9, Sept. 2015, pp. 5045–5052.

[9] "IEEE Std. 528", IEEE Standard for Inertial Sensor Terminology, 2001.

[10] M. M. Tehrani, "Ring laser gyro data analysis with cluster sampling technique," 1983 Proc. of SPIE, Bellingham, Washington, USA, pp. 207–220.

[11] E. Rubiola, "Phase noise and frequency stability in oscillators," Cambridge University Press, 2008.

[12] A. Hajimiri and T. Lee, "A general theory of phase noise in electrical oscillators," Solid-State Circuits, IEEE Journal of, vol. 33, no. 2, Feb 1998, pp. 179–194.

[13] R. Hopkins, J. Miola, et al., "The silicon oscillating accelerometer: A high-performance MEMS accelerometer for precision navigation and strategic guidance applications," Institute of Navigation National Technical Meeting (ION NTM), Jan. 2005, pp. 971–979.

[14] L. D. Landau and E. M. Lifshitz, "Resonance in nonlinear oscillators," in Mechanics, 3 ed. 1982, vol. 1, Course of Theoretical Physics, pp. 87–92.

[15] B. Razavi, "A study of phase noise in CMOS oscillators," IEEE Journal of Solid-State Circuits, vol. 31, no. 3, Mar. 1996, pp. 331–343.

[16] D. B. Leeson, "A simple model of feedback oscillator noise spectrum," Proceedings of the IEEE, vol. 54, no. 2, Feb. 1966, pp. 329–330.

[17] Y. Zhao, J. Zhao, X. Wang, et al., "A sub-μg bias-instability MEMS oscillating accelerometer with an ultra-low-noise read-out circuit in CMOS," IEEE J. Solid-State Circuits, vol. 50, no. 9, Sep. 2015, pp. 2113–2126.

[18] C. C. Enz and G. C. Temes, "Circuit techniques for reducing the effects of Op-Amp imperfections: autozeroing, correlated double sampling, and chopper stabilization," Proc. of the IEEE, vol. 84, no. 11, Nov. 1996.

[19] J. W. M. Rogers, D. Rahn and C. Plett, "A study of digital and analog automatic-amplitude control circuitry for voltage-controlled oscillators," IEEE J. Solid-State Circuits, vol. 38, no. 2, Feb. 2003, pp. 352–356.

[20] G. Han and E. Sanchez-Sinencio, "CMOS transconductance multipliers: a tutorial," IEEE Transactions on Circuits and Systems II: Analog and Digital Signal Processing, vol. 45, no. 12, Dec. 1998, pp. 1550–1563.

[21] X. Wang *et al.*, "27.2 A 1.2μg/$\sqrt{Hz}$-resolution 0.4μg-bias-instability MEMS silicon oscillating accelerometer with CMOS readout circuit," *2015 IEEE International Solid-State Circuits Conference - (ISSCC) Digest of Technical Papers*, San Francisco, CA, 2015, pp. 1–3.

[22] A. Sharma, M. F. Zaman and F. Ayazi, "A 104-dB Dynamic Range Transimpedance-Based CMOS ASIC for Tuning Fork Microgyroscopes," in IEEE Journal of Solid-State Circuits, vol. 42, no. 8, Aug. 2007, pp. 1790–1802.

[23] X. Wang *et al.*, "A 0.4μg bias instability and 1.2μg/$\sqrt{Hz}$ noise floor MEMS silicon oscillating accelerometer with CMOS readout circuit", IEEE Journal of solid-state circuits, vol. 52, no. 2, Feb. 2017, pp. 472–482.

[24] J. Zhao, Y. Zhao, X. Wang, *et al.*, "A system decomposition model for phase noise in silicon oscillating accelerometers," in IEEE Sensor Journal, vol. 16, no. 13, Jul. 2016, pp. 5259–5269.

4

An MEM Silicon Oscillating Accelerometer Employing a PLL and a Noise Shaping Frequency-to-Digital Converter

Jian Zhao[1], Yong Ping Xu[2] and Yan Su[3]

[1]Tsinghua University, Beijing, China
[2]Department of Electrical & Computer Engineering, National University of Singapore, Singapore
[3]Nanjing University of Science and Technology, Nanjing, China
E-mail: zhaojianycc@mail.tsinghua.edu.cn

This chapter presents a PLL-based readout circuit architecture for a micro-electromechanical system (MEMS) silicon oscillating accelerometer (SOA). Comparing with readout circuits based on the automatic amplitude control (AAC) technique, this new readout circuit aims to improve the bias-instability of the MEMS SOA by eliminating the dominant noise from the AAC circuit. With a modified PLL-based frequency-to-digital converter (MPLL-FDC), higher-order noise shaping is achieved, while the power consumption of the readout circuit is reduced. This chapter describes the design of the PLL-based readout circuit and presents a prototype high-performance MEMS oscillating accelerometer with the readout circuit implemented in CMOS technology.

4.1 Introduction

An MEMS accelerometer consists of an MEMS transducer and its readout circuit. The readout circuit plays an important role in high-performance accelerometers as noises originating from the readout circuit dictate the performance of the accelerometer. In MEMS SOAs, the readout circuits form two oscillators with MEMS resonators. The output of the accelerometer

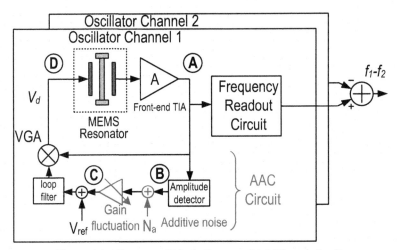

Figure 4.1   Conceptual block diagram of AAC-based circuit.

is the frequency difference between the two MEMS oscillators, as shown in Figure 4.1. Most MEMS SOAs employ an automatic amplitude control (AAC) circuit to ensure a stable oscillation amplitude, as shown in the lower part of Figure 4.1, where a variable gain amplifier (VGA) is employed in the AAC circuit to regulate the oscillation amplitude and prevent the resonator beam displacement from entering the strong nonlinear region. However, as the amplitude information in the oscillation signal is demodulated, the flicker noise in AAC blocks will contaminate the drive signal and hence the displacement amplitude. In the MEMS oscillator, as there exists the amplitude-stiffening effect (A-S effect), the nonlinear terms from the spring constant can cause a peak-frequency shift as the vibration-amplitude increases, as shown in Figure 4.2 [1]. Under such an effect, the flicker noise in the amplitude will be modulated to oscillation frequency and deteriorate the bias-instability of the MEMS SOA. Moreover, since the flicker noise from two AAC circuits in the two separate oscillator channels are not correlated, a fully differential topology of the MEMS SOA cannot reject these noises. As a result, the flicker noise in AAC increases the phase noise and deteriorates the bias-instability of the SOA.

This chapter describes a MEMS SOA with a PLL-based readout circuit, aiming to improve the bias-instability. Noise shaping frequency readout circuits are also presented. The rest of this chapter is organized as follows. Section 4.2 describes the PLL-based MEMS SOA architecture. Section 4.3

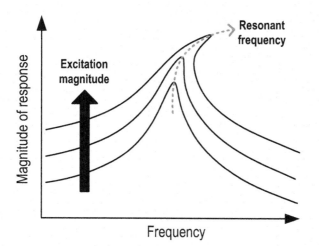

Figure 4.2   Frequency-amplitude response of a nonlinear resonator.

introduces a modified PLL-based frequency-to-digital converter (MPLL-FDC). Section 4.4 discusses the stability of the MPLL-based FDC. Section 4.5 analyses the phase noise of the overall system based on system-decomposition model. Section 4.6 deals with the circuit level implementation. The experimental results are presented in Section 4.7, followed by the conclusion in Section 4.8.

4.2 PLL-Based MEMS SOA

The concept of the PLL-based MEMS SOA CMOS readout circuit proposed in [2] is shown in Figure 4.3. In each oscillator channel, a PLL is employed to track the phase of the front-end output (oscillator output) and provides a correct phase shift for the feedback signal (drive signal) to sustain the oscillation. Meanwhile, the drive signals for both oscillators are generated by a differential low noise voltage reference (V_{ref}) in the drive signal generator, whose polarity is controlled by the feedback signal from PLL through on-chip switches. The drive amplitude is set by V_{ref} and the AAC circuit is not needed. The amplitude noise of the drive signal is dominated by the noise (N_{ref}) in V_{ref}. Since the drive signals for both oscillators are derived from the same reference, the amplitude noises induced frequency noises in two oscillator channels are correlated and hence can be rejected by the differential topology at MEMS SOA output ($f_1 - f_2$). Thus, the bias-instability can be improved. In other words, in the PLL-based readout circuit, amplitude

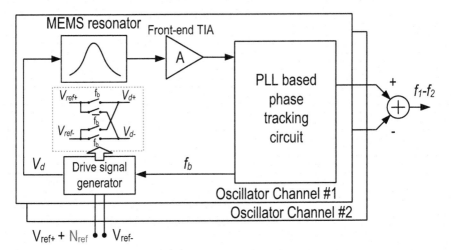

Figure 4.3   Conceptual block diagram of PLL-based MEMS SOA.

information is not demodulated to DC, like in AAC circuit, and thus avoids being contaminated by the uncorrelated flicker noise from the AAC circuit.

However, there are some issues if the readout circuit is implemented according to Figure 4.3. Firstly, there could be possible noise aliasing at the interface between the front-end and PLL, and secondly, the system (oscillator) cannot start by itself. These two issues need to be carefully addressed.

4.2.1 Noise Aliasing

Generally, the bandwidth of the front-end amplifier needs to be at least 10 times larger than the oscillation frequency in order to reduce the excess phase shift in the oscillator loop. The output of the front-end amplifier, which is also the output of the MEMS oscillator, is a sinusoidal signal containing a wideband noise. As phase detectors (PD) in the subsequent PLL effectively sample the front-end output, it may cause noise folding.

The PD can be divided into two categories, namely, the discrete time PD and continuous time PD. Discrete PD is usually edge-triggered, such as XOR PD and tri-state phase frequency detector (PFD). Since the output signal from the MEMS oscillator is sinusoidal, a waveform shaping circuit is required. Taking a tri-state PFD in Figure 4.4 as an example, as the detected phase only changes once in an input period, therefore, the noise of front-end amplifier can be regarded as being under sampled, causing noise folding and resulting in high in-band noise.

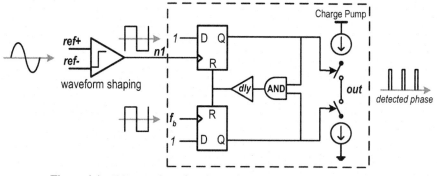

Figure 4.4 Tri-state phase frequency detector with waveform shaping.

Let us closely look at the waveform shaping circuit. As tri-state PFD only supports square wave input, the sine-wave output from the front-end must be shaped to square wave before being applied to PFD. This can be done by a comparator. As shown in Figure 4.5(a), the comparator has finite gain, and it

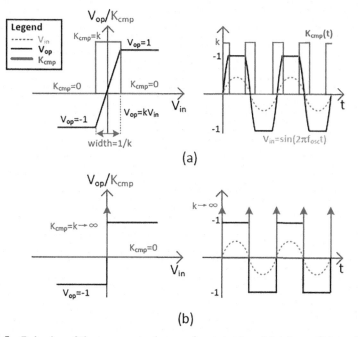

Figure 4.5 Behavior of the comparator in waveform shaping, (a) k has a finite value, and (b) k is infinity [2]. (Courtesy of IEEE).

is given by

$$K_{cmp}(t) = \begin{cases} k & |V_{op}| < 1 \\ 0 & |V_{op}| \geq 1 \end{cases} \tag{4.1}$$

where k is the comparator linear gain and V_{op} is the comparator output. The input of the comparator is directly from the oscillator output, and the response of the comparator during waveform shaping is shown in Figure 4.5(b). The gain of the comparator can be considered as a pulse function with a frequency of $2f_{osc}$, amplitude of k, and pulse width of $2/k$. If k is large enough, assuming, $k \to \infty$, the Fourier transformation of this time-variant comparator gain is

$$K_{cmp}(f) = \frac{1}{2\pi} + \frac{1}{\pi} \sum_{n=1}^{\pi} \cos(\pi n f_{osc}t) \tag{4.2}$$

Equation (4.2) implies that the comparator for waveform shaping operates like a sampling circuit in the time domain. The noise in the vicinity of all even harmonics of f_{osc} will be folded to near DC, causing phase error during waveform shaping, as shown in Figure 4.6.

To deal with this problem, an anti-aliasing band-pass filter can be added at the output of the front-end amplifier. However, the resonant frequency of

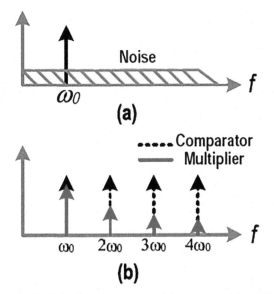

Figure 4.6   Noise aliasing behavior, (a) spectrum of sine wave with band-limited white noise, (b) spectra of comparator and multiplier gain.

an MEMS resonator is typically several tens of kilohertz. Implementing a band-pass filter on chip could be area consuming.

Another option is to use a continuous-time phase detector, i.e. an analog multiplier, which provides continuous-time phase detection and does not need waveform shaping. In contrast, as feedback signal of PLL is usually square wave, the equivalent gain function of the multiplier is given by

$$K_{mul}(f) = \frac{1}{2\pi} + \sum_{n=1}^{\infty} \frac{\sin(n\pi/2)}{n\pi/2} \cos(2\pi n f_{in} t) \qquad (4.3)$$

which is different from Equation (4.2). The harmonics in the gain function of the multiplier are modified by a sinc function whose magnitude decreases with frequency, as shown in Figure 4.6. This implies that the influence of noise folding is significantly attenuated.

4.2.2 Start-up Issue

At start-up when power is turned on, PLL cannot immediately produce the correct phase for feedback drive signal as the oscillation has not started. To start the oscillation, a start-up circuit is required. At the start-up phase, a start-up circuit based on the conventional AAC, similar to those in [3, 4], provides a feedback signal with correct phase and amplitude to drive the MEMS resonator and start oscillation. When the oscillation is established, the PLL is able to lock to the correct phase of the oscillator output and then takes over to sustain the oscillation, while the start-up circuit will be turned off to save power. Figure 4.7 shows the simplified block diagram of the PLL-based MEMS SOA, including the start-up circuit.

4.2.3 PLL Phase Tracking

As illustrated in Section 4.2.1, employing an analog multiplier as a phase detector in PLL can avoid noise folding. However, as shown in Figure 4.8, due to the sinusoidal characteristics, the multiplier-based phase detector has a limited frequency acquisition range [5]. Moreover, it cannot distinguish the harmonics of an input waveform and may falsely lock to a wrong frequency. Therefore, lock-in assistance is needed to guarantee correct frequency locking. By contrast, a tri-state PFD does not have such a problem as it only has one equilibrium at zero phase point, as shown in Figure 4.9. Taking the advantages of both phase detectors, it can avoid noise aliasing, while still lock to the correct phase.

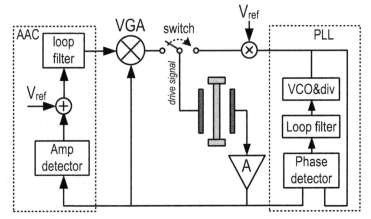

Figure 4.7   Simplified block diagram of the PLL-based MEMS SOA with the start-up circuit.

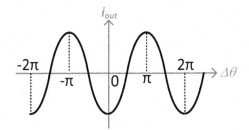

Figure 4.8   Transfer characteristic of an analogue multiplier.

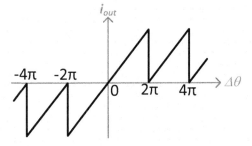

Figure 4.9   Transfer characteristic of a tri-state phase frequency detector.

A hybrid PFD shown in Figure 4.10 [2] combines a tri-state PFD with a large dead zone [5] and an analog phase detector. The tri-state PFD with a large dead zone is used to guarantee that the PLL only locks to the oscillation frequency and the analog phase detector tracks the exact phase. In other

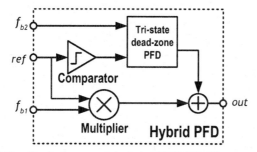

Figure 4.10 Block diagram of the hybrid PFD [2]. (Courtesy of IEEE).

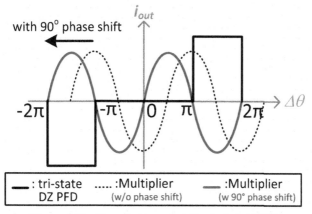

Figure 4.11 Transfer characteristic of a hybrid PFD.

words, tri-state dead-zone PFD does the coarse detection and the analog phase detector provides fine phase tracking while avoiding the noise aliasing.

Figure 4.11 shows the superimposed characteristics of analog, tri-state, and hybrid phase detectors. A 90° phase shift is added in the multiplier feedback loop to realize a symmetrical response. The 90° phase shift can be readily generated by the divider in the PLL.

4.3 PLL-Based Sigma-Delta FDC

MEMS SOA is based on the force sensing principle with frequency modulation (FM), in which the input acceleration is modulated onto MEMS oscillator frequency. Its output is effectively a time-domain signal. While the time-domain signal processing has many advantages, to facilitate interface with other circuits or devices, it is convenient to convert the output of the

MEMS SOA to a standard digital format. While a straightforward solution is to digitize the output with a conventional high-resolution analog-to-digital converter (ADC), a dedicate FDC may provide sufficient resolution at a much lower power consumption. This section gives a brief review of the existing FDCs and then introduces a PLL-based noise-shaping FDC.

4.3.1 A Brief Review of Existing FDCs

4.3.1.1 Reset counter-based FDC

A block diagram of a typical reset-counter-based FDC and its waveforms are shown in Figures 4.12 and 4.13, respectively [6]. The input to the FDC is a square-wave shaped output from the MEMS oscillator. The frequency of the input, f_{in}, may be same or higher than the original oscillation frequency. The counter accumulates the clock cycles, and the phase quantization is performed at the rising edge of the input at f_{in} at the first register. The output, $D[n]$, is the difference between two adjacent counter outputs, $C[n]-C[n-1]$, which represents the phase difference of two adjacent f_{in} cycles, that is,

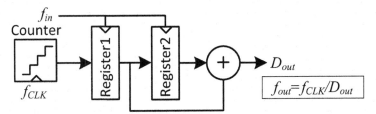

Figure 4.12 Block diagram of the reset-counter-based FDC.

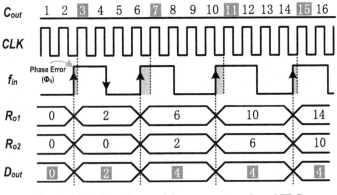

Figure 4.13 Waveforms of the reset-counter-based FDC.

the instantaneous period of the input. Thus instantaneous f_{in} can be found by $f_{in} = 1/(T_{CLK}D_{out}) = f_{CLK}/D_{out}$. The Register 2 and Subtractor also form a differentiator that provides a first-order noise shaping to the phase quantization error, Φ_q, which is given by

$$D_{out}[n] = \frac{\Phi_{CLK}(z) + (z^{-1} - 1)\,\Phi_q(z)}{2\pi} \tag{4.4}$$

The phase quantization noise is therefore attenuated at a low frequency. The reset counter FDC is very energy efficient and has been employed in several oscillator-based sensors [3, 7, 8]. However, first-order noise shaping has to rely on high clock frequency to achieve the desired resolution. A reset counter FDC in [6] achieves 71.8 dB SNR in 100 kHz bandwidth with a 500 MHz clock, an oversample ratio of 1000.

4.3.1.2 Delta-sigma FDC ($\Sigma\Delta$ FDC)

Figure 4.14 shows a sigma-delta ($\Sigma\Delta$) modulation-based FDC reported in [9, 10]. The $\Sigma\Delta$ FDC can be considered as a phase-domain $\Sigma\Delta$ ADC, where the input is phase and an ADC is inserted after the loop filter to digitize the phase error. The output of the ADC, D_{out}, is converted back to the phase domain through a digitally controlled oscillator (DCO) formed by a reset counter and a reference clock, and fed back to the input of the phase detector. The counter is reset according to D_{out}, i.e., the counter accumulates the reference clock and resets when it reaches D_{out}. The phase detector extracts the phase error between zero-crossing of the input and the reset-edge of the counter output. With a sufficient loop gain, the phase error at input approaches zero and D_{out} represents the instantaneous input frequency. The quantization error, q_n, from ADC is shaped by the loop filter, and the FDC output, D_{out}, is

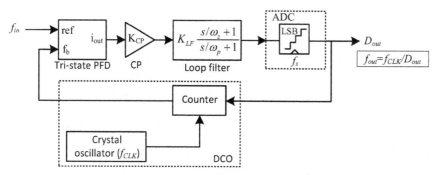

Figure 4.14   Sigma-delta FDC, with second order quantization noise shaping.

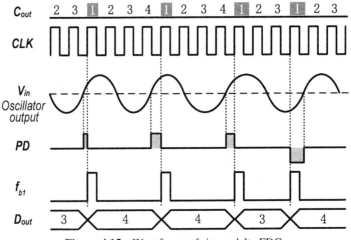

Figure 4.15   Waveforms of sigma-delta FDC.

given by

$$D_{out} = \left(\frac{G_L}{1 + G_L}\right) \frac{f_{CLK}}{2f_{in}} + \frac{q_n}{1 + G_L} \qquad (4.5)$$

where G_L is the loop gain. The waveforms in Figure 4.15 illustrate the operation of the FDC, where f_{in} represents the zero-crossing waveform of the sinusoidal input signal from the oscillator. As discussed in Section 4.2.1, the zero-crossing detection can be considered as a sampling process at a sampling frequency of $2f_{in}$ (i.e. twice per input cycle). Since the oscillator output contains wideband noise, the zero-crossing detection introduces noise aliasing and limits the resolution of the FDC. Furthermore, to achieve a high-order noise shaping, the loop filter design may not be straightforward for the loop stability.

4.3.1.3 PLL-based FDC (PLL-FDC)

Figure 4.16 shows a PLL-based FDC reported in [11]. The architecture is similar to a conventional PLL, but with two D-Flip Flops (DFF) added at the output, which extracts the input frequency and, at the same time, provides a first-order noise shaping by including the first DFF inside PLL loop. Together with the loop filter and VCO, an overall third-order phase quantization noise shaping is achieved with a low hardware budget. The PLL-FDC only has 1 bit output at clock frequency, f_{CLK}, and needs to be decimated to obtain the final frequency measurement result. A waveform example in Figure 4.17 illustrates

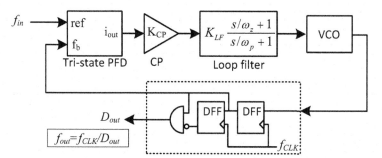

Figure 4.16   PLL-based FDC with third order quantization noise shaping.

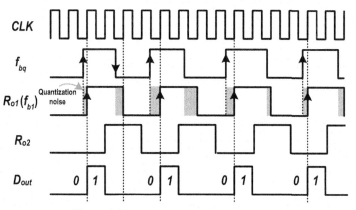

Figure 4.17   Waveform example of PLL-FDC.

the operation of the PLL-FDC. A transistor-level simulation has shown that a 78 dB signal-to-noise and distortion ratio (SNDR) can be achieved with a 212.7 MHz sampling clock [12].

4.3.2 A Modified PLL-based FDC (MPLL-FDC)

In the PLL-based MEMS SOA described in this chapter, a modified PLL-FDC (MPLL-FDC) is employed [13], as shown in Figure 4.18. The MPLL-FDC incorporates a reset-counter FDC in the PLL-FDC. A DFF is inserted after the divider in the PLL feedback loop. It retimes signal, f_{bq}, making it synchronized with the reset counter clock (CLK). Since the DFF is inside the PLL loop, the phase error introduced by the retiming is second-order shaped by the PLL loop. The retimed signal, f_{b1}, is used to capture the counter output through Register 1, which effectively reflect its phase. Subsequent Register 2

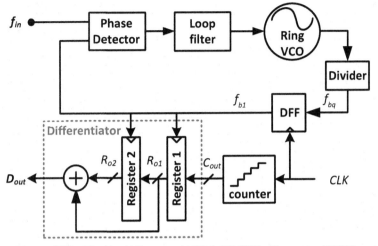

Figure 4.18   The block diagram of MPLL-FDC [2]. (Courtesy of IEEE).

and the subtractor form a differentiator which extracts the instantaneous cycle of f_{b1}, with first-order noise shaping, similar to the DFFs in PLL-FDC. Note that $f_{b1}=f_{in}$ when the PLL is locked. Since f_{b1} and CLK are now synchronized, there is no phase error introduced when the counter output is sampled by Register 1. Thus, the remaining error in f_{b1} is further first-order shaped by the differentiator. In other words, together with the differentiator, the phase quantization error is effectively third-order shaped in digitized D_{out}. When PLL is locked, D_{out} can expressed as:

$$D_{out} = \frac{T_{in}}{T_{CLK}} \tag{4.6}$$

Thus, the input frequency can be obtained,

$$f_{in} = \frac{f_{CLK}}{D_{out}} \tag{4.7}$$

It should be mentioned here that the clock frequency should be at least 2x higher than the feedback signal (f_{bq}) so that the retiming would not miss any edges in f_{bq}. The timing diagram of the MPLL-FDC is shown in Figure 4.19. The detailed noise analysis will be discussed in the next subsection. The main difference between MPLL-FDC and PLL-FDC is that no decimation is required for MPLL-FDC and the input frequency can be directly obtained from FDC output.

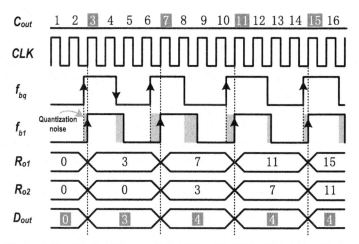

Figure 4.19 Timing diagram of the MPLL-FDC [2]. (Courtesy of IEEE).

4.3.3 Analysis of Quantization Error in MPLL-FDC

To analyze the frequency resolution, a simplified linear model of MPLL-FDC is shown in Figure 4.20, where k_d is the gain of the phase detector. $L(s)$ is the transfer function of the loop-filter. Thus, the signal transfer function (STF) and the noise transfer function (NTF) can be obtained as follows:

$$STF(s) = \frac{F_{out}(s)}{F_{in}(s)} = \frac{k_d k_{vco} L(s)}{s + k_d k_{vco} L(s)} \tag{4.8}$$

$$NTF(s) = \frac{F_{out}(s)}{N_q(s)} = \frac{s^2}{2\pi [s + k_d k_{vco} L(s)]} \tag{4.9}$$

where k_{vco} is the gain of VCO. If $L(s)$ is a Type-II loop filter, the $NTF(s)$ will provide third-order high-pass noise shaping.

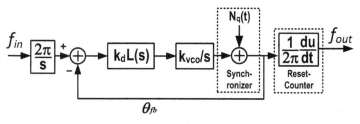

Figure 4.20 Linearized model of MPLL-FDC.

The noise source N_q is a phase error introduced by the retiming of f_{bq}. Since the retiming error is within $[0, T_{CLK}]$, the phase error is distributed within $[0, 2\pi f_{in} T_{CLK}]$. Since there is no correlation between the sampling clock and the VCO, this phase quantization error can be regarded as uniformly distributed from $-(2\pi f_{in} T_{CLK})/2$ to $(2\pi f_{in} T_{CLK})/2$ with the density of $1/(2\pi f_{in} T_{CLK})$ [6], and can be expressed as

$$N_q = \frac{2\pi f_{in} T_{CLK}}{\sqrt{12}} \tag{4.10}$$

The phase quantization error is inversely proportional to the clock frequency, f_{CLK}. The phase quantization noise power spectrum density in f_{out} can be expressed as:

$$N_{out}^2(f) = \frac{(2\pi f_{in} T_{CLK})^2}{12} |NTF(2\pi f)|^2 \tag{4.11}$$

For a given band-width of f_{BW}, the average in-band noise power density is given by:

$$P_{IBN} = \frac{(2\pi f_{in})^2}{12 f_{CLK}^2 f_{BW}} \int_0^{f_{BW}} |NTF(2\pi f)|^2 \, df \quad [\text{radian}^2/\text{Hz}] \tag{4.12}$$

Figure 4.21 shows the simulation results in SIMULINK and the parameters used in the simulation configurations are listed in Table 4.1. The bandwidth of PLL is set to be 2 kHz. Based on the transient simulation results,

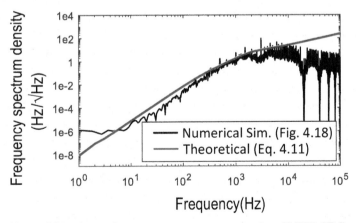

Figure 4.21   Output frequency power spectrum density of MPLL-FDC.

the power spectrum density of f_{out} can be obtained. Figure 4.21 also compares the simulated result with the theoretical prediction by Equation (4.12), in which the third-order quantization noise shaping effect can be observed. In addition, the simulated noise spectrum density agrees well with that from theoretical prediction.

Figures 4.22 and 4.23 compare the simulated output power spectrum densities of the four FDCs. During the simulation, the $\Sigma\Delta$-FDC employs a 5-bit ADC as the embedded voltage quantizer (to quantize the control voltage of DCO). The other parameters are same as those in Table 4.1.

It can be seen from the simulation results in Figure 4.22 that both MPLL-FDC and PLL-FDC achieve third-order noise shaping, while the $\Sigma\Delta$-FDC

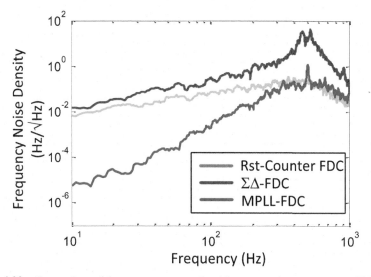

Figure 4.22 Comparison of the power spectrum densities among the reset-counter, $\Sigma\Delta$-FDC and MPLL-FDC.

Table 4.1 Simulation parameters

Parameters	Values	Description
f_{in}	23 kHz	Input frequency
f_{CLK}	200 kHz	Counting frequency
G_L	1.2e7	Loop gain of PLL, which is the product of K_{VCO}, K_d, K_{lf}
f_{vco}	1 kHz	Center frequency of VCO
ω_z	1.5 krad/s	Zero frequency of loop filter
ω_p	20 krad/s	Pole frequency of loop filter

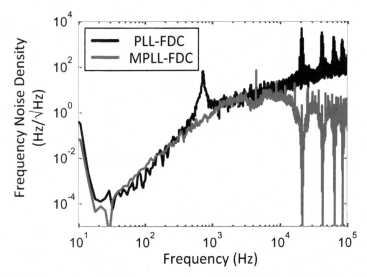

Figure 4.23 Comparison of the power spectrum density between PLL-FDC and MPLL-FDC.

and reset counter FDC only provides second-order and first-order noise shaping, respectively. Moreover, compared to the $\Sigma\Delta$-FDC in Figure 4.23, the MPLL-FDC can attenuated the high-frequency noise due to its inserted counter.

4.4 Stability of the PLL with a Hybrid PFD

The block diagram of the MPLL-FDC is shown in Figure 4.24, where f_{b1} is fed back to an analog phase detector, with an additional 90° phase delay, and f_{b2} goes to a tri-state dead-zone PFD. The feedback signal, f_{b1}, also controls the driving signal phase through a set of switches. The loop filter is a typical Type-II filter with a buffer driving a ring oscillator-based VCO. The quantizer in the MPLL-FDC is described in the previous section.

Figure 4.25 shows a system model of the PLL in Figure 4.24, where $L(s)$ is the loop filter transfer function, K_{mul} is the multiplier gain, K_{CP} is the gain of a charge pump in the tri-state dead-zone PFD. $f_1(\Delta\theta)$ and $f_2(\Delta\theta)$ represent the functions of the analog multiplier and tri-state dead-zone PFD, respectively, and are given by

$$f_1(\Delta\theta) = \frac{2AB}{\pi}\sin(\Delta\theta) \qquad (4.13)$$

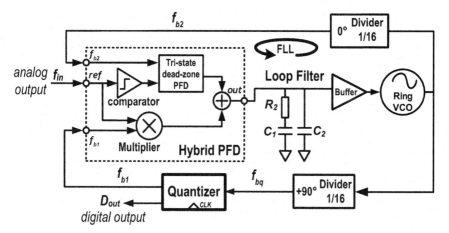

Figure 4.24 Block diagram of MPLL-FDC employing hybrid PFD.

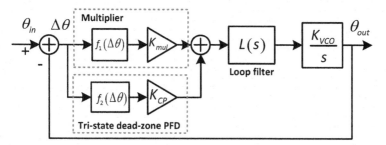

Figure 4.25 Model of the PLL with the hybrid PFD [2]. (Courtesy of IEEE).

$$f_2\left(\Delta\theta\right) = \begin{cases} 0 & \left(2n+1\right)\pi < |\Delta\theta| \leq \left(2n+2\right)\pi \\ \mathrm{sgn}\left(\Delta\theta\right) & 2n\pi < |\Delta\theta| \leq \left(2n+1\right)\pi \end{cases} \quad (n = 0, 1, 2 \ldots n)$$

(4.14)

where A and B are the amplitudes of two input signals to the analog phase detector. Here a sine-wave reference and a square-wave feedback signal are assumed.

The overall function of hybrid PFD, including $K_{mul}f_1$ and $K_{CP}f_2$, are plotted in Figures 4.26 to 4.28 with different gain ratio, $\gamma = K_{CP}f_2/(K_{mul}f_1)$ $=\pi K_{CP}/(2ABK_{mul})$. Within $[-2\pi:2\pi]$ range, the response of dead-zone PFD can be regarded as a sign function with $[-\pi:\pi]$ dead zone [5]. Intuitively, the gain ratio will determine which sub-phase detector contributes more to the overall function of the hybrid PFD.

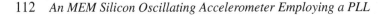

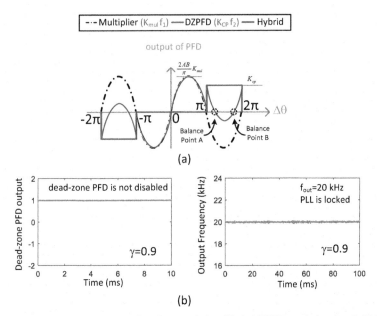

Figure 4.26   For $\gamma < 1$, (a) Transfer characteristic of hybrid PFD, and simulated (b) hybrid PFD output, and (c) frequency of PLL output.

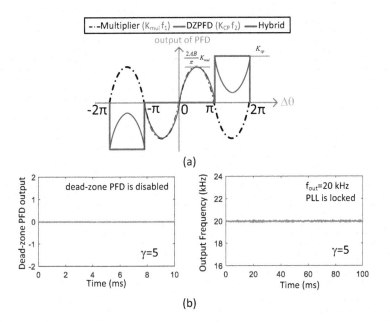

Figure 4.27   For $\gamma > 1$, (a) Transfer characteristic of hybrid PFD, and simulated (b) output of hybrid PFD, and (c) frequency of PLL output.

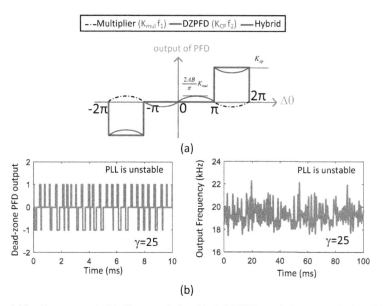

(a)

(b)

Figure 4.28 For $\gamma >> 1$, (a) Characteristic of hybrid PFD, and simulated (b) hybrid PFD output, and (c) frequency of PLL output.

Figure 4.26 is a simulation result for $\gamma < 1$, where input signal frequency is 20 kHz. It can be seen that the output of the hybrid PFD reaches zero not only at zero phase, but also other phases. For example, in the $[\pi:2\pi]$ region, there are two additional zero-crossing points, N and P. Among them, point N has negative derivation which is unstable, while point P has positive derivation and is a stable point. In this case, the PLL locks to a correct frequency as shown in Figure 4.26(b), but a wrong phase. Furthermore, since the zero-crossing point P is not in the dead zone of tri-state PFD, it implies that the tri-state PFD will never be disabled since it is outside of its dead-zone as shown in Figure 4.26(a). In other words, the analog phase detector will never be activated, thus resulting in a large phase error although the PLL is locked.

For $\gamma > 1$, as shown in Figure 4.27, there is only one stable lock point at $\Delta\theta = 0$. This implies that the PLL will lock to both correct frequency and phase. It should be noted that γ cannot be unbounded large, because if γ is very large, the hybrid PFD will behave like a dead-zone PFD only. A so-called "chattering" behavior will be triggered due to the enlarged dead-zone of PFD [14]. As a result, the PLL will fail to lock. This can be seen in Figure 4.28, where $\gamma = 25 >> 1$. The dead-zone PFD takes over and dominates the response. Due to its large dead-zone, the output of dead-zone PFD is unstable as shown in Figure 4.28(a), thus the PLL also cannot lock to the correct frequency as shown in Figure 4.28(b), which means the "chattering" behavior has been triggered.

4.5 Noises in PLL-Based MEMS SOA

Figure 4.29 shows a block diagram of the MEMS SOA. As discussed in Section 4.2, the PLL-based MEMS SOA readout circuit eliminates the AAC circuit. As a result, its amplitude path is an open loop. Thus, it is important to understand the phase noise characteristics of the MEMS SOA. Based on the noise decomposition method in [15] and Chapter 5 of this book, the noise model of the MEMS SOA can be decomposed into amplitude ($H_m{}^{<a>}$) and phase ($H_m{}^{<p>}$) subsystems, as shown in Figure 4.30. MEMS resonator behaves like a first-order low-pass system in the decomposed amplitude sub-systems [16], and the cut-off frequency is ω_m, which is equal to $\omega_0/(2Q)$. K_{bd} is the conversion gain from drive voltage to drive force of the MEMS resonator, which depends on bias voltage, V_b, and drive electrode capacitance. K_m is the mechanical gain from drive force to displacement at oscillation frequency ω_0, which is equal to $Q_d/(m_d\,\omega_0)$. The decomposed noise sources

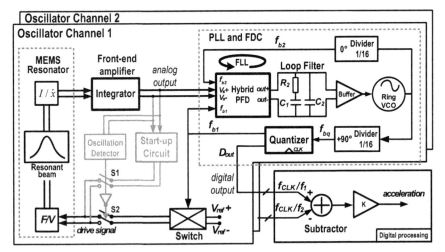

Figure 4.29 Overall system block diagram of PLL-based MEMS SOA [2]. (Courtesy of IEEE).

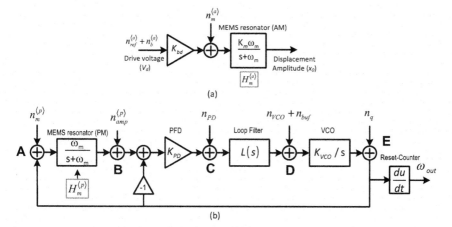

Figure 4.30 (a) Amplitude sub-system and (b) Phase sub-system of proposed MEMS SOA for noise analysis [2]. (Courtesy of IEEE).

Table 4.2 Definitions of noise sources

Noise	Catagory	Remarks
n_m	additive	Mechanical noise along drive direction
n_{amp}	additive	Input referred noise of front-end amplifier
n_{PD}	additive	Input referred noise of phase detector
n_{VCO}	additive	Output noise of VCO
n_q	additive	Quantization noise
n_{buf}	additive	Output voltage noise of D2S buffer
n_b	multiplicative, stiffness modulation	Noise of bias voltage on MEMS electrode
n_{ref}	multiplicative	Noise of reference voltage for amplitude control

are also shown in Figure 4.30. The definitions of the noise sources are listed in Table 4.2.

Figure 4.30(a) is the amplitude subsystem. x_0 is only controlled by the reference voltage from the drive signal generator. As a result, only the reference noise, n_{ref}, mechanical thermal noise, n_m, and bias voltage noise, n_b, need to be considered. Among them, flicker noises in n_{ref} and n_b directly modulate the displacement amplitude and thus degrade the bias-instability. Fortunately, as two oscillator channels share the same reference and bias, there noises, n_{ref} and n_b, can be considered as common-mode interferences and cancelled at the final SOA output. However, in practice, due to the mismatches between the two oscillator channels after fabrication, the noises from amplitude subsystem cannot be completely cancelled. Some calibration may be needed.

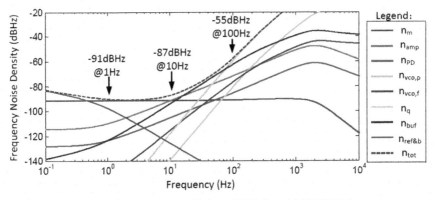

Figure 4.31   Noise prediction of PLL-based MEMS SOA.

Figure 4.30(b) shows the detailed phase sub-system for noise analysis. In the PLL, the VCO noise and the phase quantization noise are second-order high-pass shaped. Thus, the overall noise at low frequency is significantly attenuated, and hence improves the bias-instability and resolution. But the elevated noise at a higher frequency may limit the effective bandwidth of the MEMS SOA.

The simulation results based on the parameters in Table 4.2 are shown in Figure 4.31. The frequency noise floor is −91 dBHz/Hz, which is equivalent to 28 μHz/$\sqrt{\text{Hz}}$. Assume the scale factor of each resonator is 100 Hz/g; this equivalent acceleration noise floor is 0.28 μg/$\sqrt{\text{Hz}}$. This figure also shows the noise contributions from each individual noise source. Based on the system-decomposition model, the specifications for circuit design can be determined.

4.6 Key Circuit Designs for PLL-based MEMS SOA

4.6.1 Analog Front-end Amplifier

In AAC-based MEMS SOA readout circuit, the front-end amplifier not only needs to amplify the transducer signal, but also to provide correct phase shift to satisfy the oscillation condition. In the PLL-based MEMS SOA, the PLL can conveniently provide additional phase shift to satisfy the oscillation condition. Thus, the phase of the front-end output is not critical and the requirement can be relaxed. Either a charge-sensing amplifier (CSA) or a trans-impedance amplifier (TIA) can be employed.

In this design, a simple CSA is chosen for its ultralow input current noise and power consumption. The schematic of the CSA is shown in Figure 4.32(a), where C_p is the lumped parasitic capacitance between the

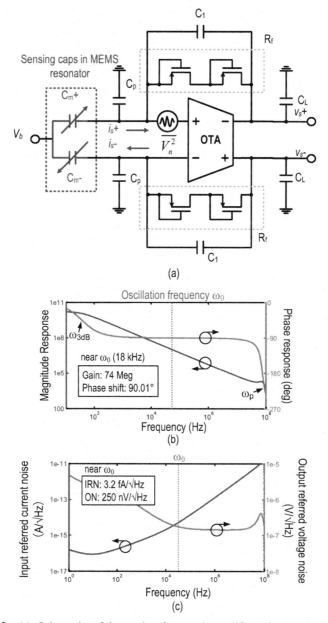

Figure 4.32 (a) Schematic of low-noise front-end amplifier, simulated (b) frequency response, and (c) noise spectrum density of front-end amplifier, where C_p denotes the parasitic capacitance at input node of OTAs [2]. (Courtesy of IEEE).

input terminals and ground, C_m is the sense electrode capacitance of the MEMS resonator. Two pseudo-resistors (R_b) are used to set the input DC bias. The input-referred current noise from these pseudo-resistors can be ignored since their resistance is extremely high, in an order of $\sim 10^{12}$ Ω or higher. The CSA is essentially a charge integrator that integrates the motion current onto feedback capacitance C_1. Figure 4.32(b) shows the simulated frequency response. The 3-dB corner and non-dominant pole frequencies (ω_{3dB} to ω_p) are determined by $(R_b C_1)^{-1}$ and bandwidth of the operational transconductance amplifier (OTA), respectively, and should be separated far away from the signal frequency (20 kHz) so that the phase shift at signal frequency is approximately 90°. Since the current noise from the pseudo-resistor is negligible, the input-referred current noise of the front-end amplifier mainly comes from the OTA. Therefore, the OTA noise needs to be minimized. The simulated input-referred noise and output noise of front-end amplifier are shown in Figure 4.32(c). The design parameters and simulated performance of the front-end amplifier are listed in Table 4.3.

4.6.2 Hybrid Mode Phase Frequency Detector

As discussed in Section 4.2.3, the hybrid PFD consists of a comparator with hysteresis, a tri-state PFD with large dead zone and an analog multiplier as the phase detector. In order to avoid multiple zero-crossing points during waveform conversion, the comparator in hybrid PFD is designed with a hysteresis and is shown in Figure 4.33.

Figure 4.34 shows the schematic of tri-state dead-zone PFD. It is implemented by a conventional tri-state PFD with a dead-zone generator. The dead-zone generator is another pair of DFFs, which is sampled by the inverted inputs. As an inverter is equivalent to a 180° delay, when the phase difference between two inputs $n1$ and f_{b2} is less than π, the output of PFD is zero,

Table 4.3   Specifications of an analog front-end amplifier

Specifications	Value
$R_b \setminus C_1 \setminus C_p$	20 GΩ \ 0.1 pF \ 1$\sim$4 pF
C_m(quiescent)	1 pF
Bias voltage V_b	10 V
Gain @ 20 kHz	64 M
Input referred current noise @ 20 kHz	3.2 fA/$\sqrt{Hz}$
Input referred capitance noise @ 20 kHz	2.5 zF/$\sqrt{Hz}$
Supply voltage	1.5 V
Total power consumption	410 μW

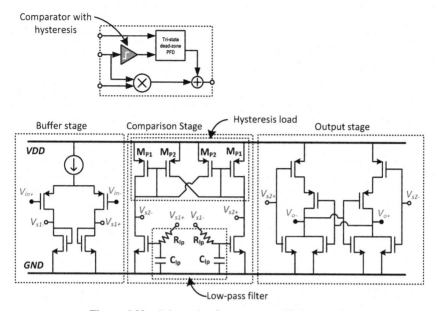

Figure 4.33 Schematic of comparator with hysteresis.

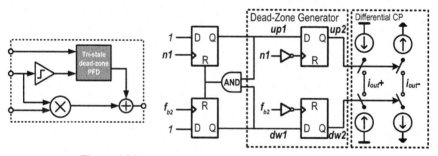

Figure 4.34 Schematic of tri-state PFD with a dead zone.

which means an $[-\pi, \pi]$ dead zone is generated. Figures 4.35 and 4.36 show a comparison between the tri-state PFD with or without dead-zone generator, in which $up1$ and $up2$ are the outputs before or after the dead-zone generator, respectively. Finally, the outputs of the analog multiplier and tri-state dead-zone PFD are connected together as the output of the hybrid PFD. Once the PLL is locked, the phase error between two inputs of PFD is close to zero and the tri-state PFD enters its dead zone.

The analog multiplier is shown in Figure 4.38. It is implemented by four analog switches preceded by a band-pass V-I converter. In such an

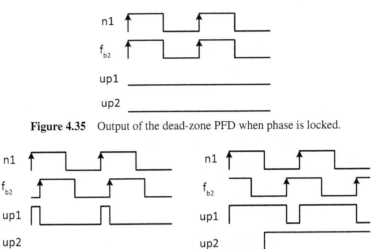

Figure 4.35 Output of the dead-zone PFD when phase is locked.

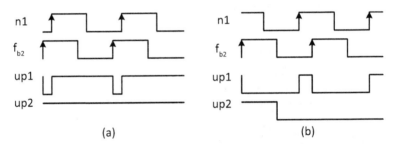

Figure 4.36 Output of the dead-zone PFD when $n1$ leads f_{b2}: (a) leading phase $< \pi$ and (b) leading phase $> \pi$.

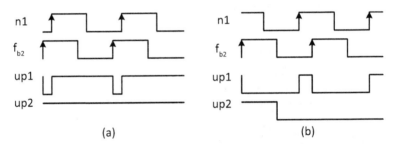

Figure 4.37 Output of the dead-zone PFD when $n1$ lags f_{b2}: (a) lagging phase $< \pi$ and (b) lagging phase $> \pi$.

implementation, band-pass V-I converter, enclosed in the dotted rectangle, filters the additive flicker noise and attenuate the multiplicative flicker noise through its close-loop topology [17]. The close-loop topology boosts the gm of MN_1 and MN_2, which makes the overall transconductance of the V-I converter equal to $C_1/(R_sC_2)$, independent of transistor parameter variations. The key design parameters of the hybrid PFD are listed in Table 4.4. Note that both analog multiplier and dead-zone PFD are fully differential designs with a common-mode feedback.

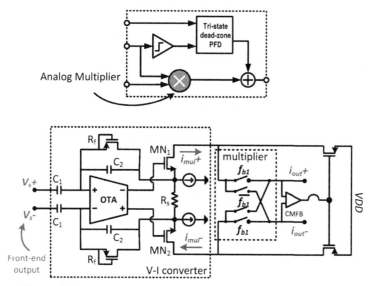

Figure 4.38 Schematic of analog multiplier [2]. (Courtesy of IEEE).

Table 4.4 Specifications of hybrid PFD

Specifications	Value
$R_b \backslash C_1 \backslash C_2 \backslash R_s$	20 GΩ\0.3 pF\0.6 pF\20 kΩ
Gain @ 20 kHz	25 μA/V
Passing band	20 Hz 2 MHz
Input referred current noise @ 20 kHz	300 nV/$\sqrt{Hz}$
Charge pumb current in DZ PFD	64 μA
Gain ratio of Hybrid PFD (γ)	6
Supply voltage	1.5 V
Total power consumption	120 μW

4.6.3 Phase-Lock Loop with FDC

The overall circuit schematic of the PLL and FDC is shown in Figure 4.39. An integer-N PLL is employed with a type-II passive loop filter. In order to isolate the loop filter from the VCO, a differential-to-single (D2S) converter is inserted. The D2S is implemented by a difference amplifier with a pair of voltage follower as the first stage to avoid the loading from subsequent difference amplifier, whose output DC voltage is set by V_{DC}.

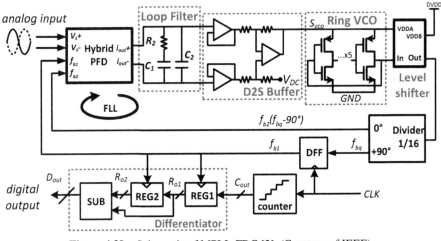

Figure 4.39   Schematic of MPLL-FDC [2]. (Courtesy of IEEE).

The VCO is a 5-stage ring oscillator, whose VDD is supplied by the D2S output and is not at a fixed value. As a result, a level shifter is required to adjust VCO output voltage level to DVDD to drive the subsequent logic blocks. The divided-by-16 frequency divider provides two orthogonal outputs, which are fed back to the tri-state dead-zone PFD and the analog multiplier, respectively, to realize a symmetrical hybrid phase response function as shown in Figure 4.11.

4.7 Experiment Results of a Prototype PLL-based MEMS SOA

4.7.1 Prototype Implementation

Figure 4.40 shows the schematic of MEMS transducer, which is fabricated in an 80−μm SOI process with a wafer-level vacuum package and sealed in a ceramic package. The die area of the MEMS transducer is 20 mm^2. The readout circuit is implemented in a standard 0.35-μm CMOS process with a chip area of 10.9 mm^2. Figure 4.41 shows the chip microphotograph and a test PCB. The readout circuit chip is mounted on the PCB through chip-on-board packaging.

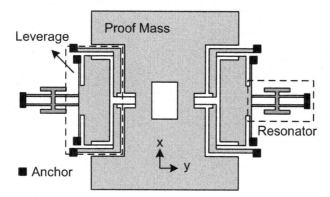

Figure 4.40 Schematic of MEMS transducer[2]. (Courtesy of IEEE Press).

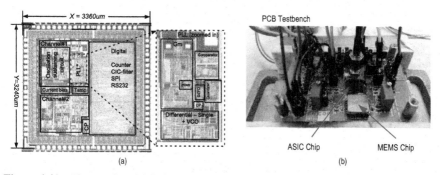

Figure 4.41 Photos of (a) fabricated readout circuit, and (b) PCB test board[2]. (Courtesy of IEEE).

4.7.2 FDC Measurement Results

To evaluate the PLL-based FDC only, the MEMS resonator and front-end amplifier is bypassed. A square-wave excitation signal is applied to the hybrid PFD, with frequency sweeping from 15 to 25 kHz, and the frequency of the counter clock is 750 kHz. Figure 4.42 shows the transfer characteristic of the FDC, the residual error within the operation range is only 20 ppb.

4.7.3 MEMS SOA Measurement Results

Figure 4.43 shows the transient waveforms measured during the start-up period. Three key signals, namely, sense, drive, and VCO control voltages,

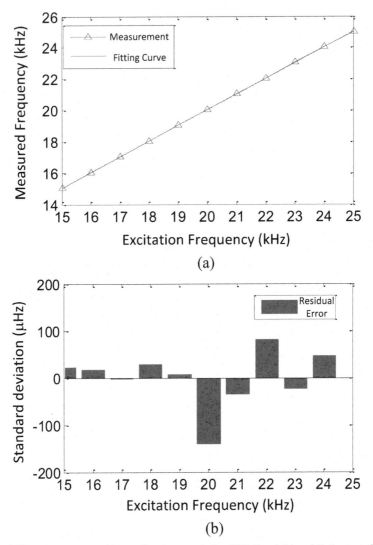

Figure 4.42   Measurement (a) transfer characteristic of FDC and (b) residual error of FDC.

are recorded by an NI-DAQ board under 100 kHz sample rate per channel. The start-up process has three phases. Before power-up (off phase), there is obviously no oscillation. After powering on, the start-up circuit provides the drive signal and the system starts the oscillation (start-up phase). During this

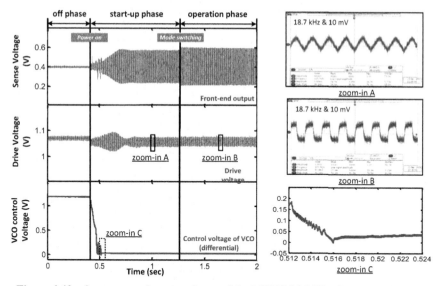

Figure 4.43 Start-up transient waveforms of the MEMS SOA[2]. (Courtesy of IEEE).

process, the PLL locks to the oscillator output frequency. When the oscillation is established, the PLL is manually switched in to take over and drive the MEMS resonator, together with drive signal generator (operation phase). The start-up circuit will then be manually turned off to save power. The manual switching between different modes is designed for prototyping. It can be easily switched automatically on chip.

To evaluate the full-scale and linearity, the accelerometer is tested on a centrifugal table. The test results show that this prototype can achieve less than 50 ppm nonlinearity within ±30 g full scale, which is shown in Figure 4.44. The output of the accelerometer is taken as the difference between two squared oscillation frequencies $(f_1^2 - f_2^2)$ from the two channels of the readout circuits [18].

Figure 4.45(a) shows the bias-stability (defined as 1-σ value of the bias error) test result. The test is carried out over three hours under room temperature, and the raw output data are recorded at a sampling rate of 732 Hz using an FPGA acquisition board and then decimated to 0.1 Hz. The acceleration error is calculated by taking the difference between two decimated output frequencies $(f_2 - f_1)$. The standard deviation calculated from one hour measurement is 0.9 µg under 0.1 Hz bandwidth, and 2.5 µg

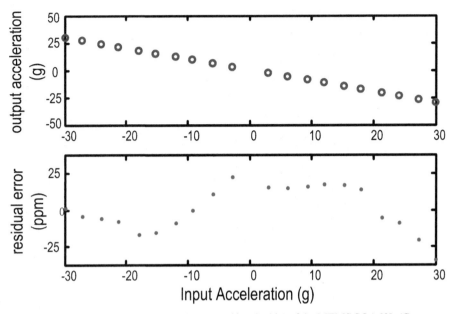

Figure 4.44   Full scale test results (with a centrifugal table) of the MEMS SOA [2]. (Courtesy of IEEE).

under 1 Hz bandwidth. The bias-instability is obtained from the same data set (without decimation) by computing its Allan variance. As shown in Figure 4.45(b), the measured bias-instability is 0.23 μg with an average time around 100 seconds.

The output acceleration noise density of the MEMS SOA is shown in Figure 4.46. The acceleration noise density of 1 μg/√Hz is achieved within 20 Hz bandwidth. It also shows that the measured PSDs agree well with those from theoretical prediction by system-decomposition model.

Taking the full scale into account, the PLL-based MEMS SOA achieves a relative instability of 4 ppb (bias instability/full input scale) and the relative noise density of 17 ppb/Hz$^{1/2}$(noise PSD/full input scale). The complete MEMS SOA, excluding the external reference, consumes only 2.7 mW under a 1.5 V supply. Table 4.5 compares the PLL-based MEMS SOA with prior art. Figure 4.47 shows the comparisons of relative bias instability and acceleration noise density of the recently published MEMS accelerometers, including both MEMS SOA and capacitive accelerometers.

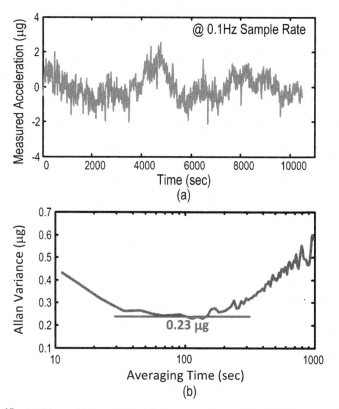

Figure 4.45 (a) Bias stability and (b) Allan variance test results [2]. (Courtesy of IEEE).

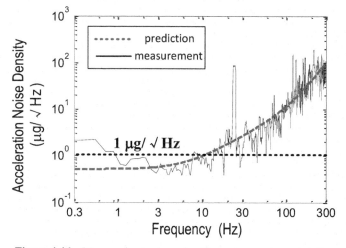

Figure 4.46 Measured output acceleration noise spectrum density.

Table 4.5   Performance comparison

Reference	[19]	[20]	[21]	[4]	[3]	This work
Year	2012	2015	2015	2008	2015	2016
Mechanism	Capacitive	Capacitive	Capacitive	SOA	SOA	SOA
Process (μm)	0.35	N.A.	0.5	0.35	0.35	0.35
Supply (V)	3.6	N.A.	7	3.3	1.5	1.5
Full scale (g)	±1.15	±15	±1.2	±20	±20	±30
Power (mW)	3.6	400	23	23	4.4	2.7
Bias instability (μg)	13	0.8	18	4	0.4	0.23
Noise density ($\mu g/\sqrt{Hz}$)	2	1	0.2	20	1.2	1
Relative instability (ppb)	5650	27	7500	100	10	4
Relative noise density (ppb/$\sqrt{Hz}$)	870	33	83	500	30	17
Readout circuit	Σ-Δ ADC	Σ-Δ ADC	Σ-Δ ADC	AAC based	AAC based	PLL based

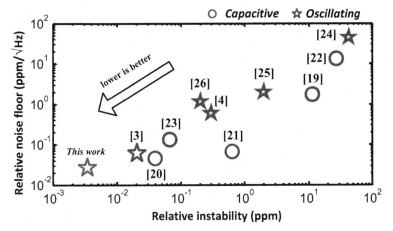

Figure 4.47   Comparisons of relative bias-instability and noise density [2]. (Courtesy of IEEE).

4.8 Conclusion

This chapter introduces a PLL-based readout circuit for MEMS SOA and demonstrates its potential to achieve high performance with low power consumption. The PLL is used to provide the correct phase for the MEMS oscillator and a low-noise external reference sets the required feedback drive signal amplitude. A hybrid PFD is employed in the PLL to avoid noise folding during phase acquisition. Furthermore, a frequency-to-digital converter, partially embedded in the PLL loop, is able to achieve third-order quantization noise shaping with low power consumption. Finally, a PLL-based MEMS SOA prototype with CMOS readout circuit has been demonstrated and achieved state-of-the-art performance.

Acknowledgement

The author would like to thank Yang Zhao and Xi Wang for the insightful discussions and assistance during the circuit design. The author also want to thank Jinan Wang for the his supports during sensor testing.

References

[1] L. H. L. He, Y. P. X. Y. P. Xu, and M. Palaniapan, "A State-Space Phase-Noise Model for Nonlinear MEMS Oscillators Employing Automatic Amplitude Control," *IEEE Trans. Circuits Syst. I Regul. Pap.*, vol. 57, no. 1, pp. 189–199, 2010.

[2] J. Zhao, X. Wang, Y. Zhao, G. M. Xia, A. P. Qiu, Y. Su, and Y. P. Xu, "A 0.23-μg Bias Instability and 1-μg/$\sqrt{Hz}$ Acceleration Noise Density Silicon Oscillating Accelerometer With Embedded Frequency-to-Digital Converter in PLL," *IEEE J. Solid-State Circuits*, vol. 52, no. 4, pp. 1053–1065, Apr. 2017.

[3] X. Wang, J. Zhao, Y. Zhao, G. M. Xia, A. P. Qiu, Y. Su, and Y. P. Xu, "A1.2 μg/$\sqrt{Hz}$-resolution 0.4 μg-bias-instability MEMS silicon oscillating accelerometer with CMOS readout circuit," *Dig. Tech. Pap. - IEEE Int. Solid-State Circuits Conf., vol. 58, pp.* 476–477, 2015.

[4] L. He, Y. P. Xu, and M. Palaniapan, "A CMOS Readout Circuit for SOI Resonant Accelerometer With 4-Bias Stability and 20-Resolution," Solid-State Circuits, IEEE J., vol. 43, no. 6, pp. 1480–1490, 2008.

[5] X. Gao, E. A. M. Klumperink, M. Bohsali, and B. Nauta, "A low noise sub-sampling PLL in which divider noise is eliminated and PD/CP noise is not multiplied by N^2," *IEEE J. Solid-State Circuits*, vol. 44, no. 12, pp. 3253–3263, 2009.

[6] J. Kim, T. K. Jang, Y. G. Yoon, and S. Cho, "Analysis and design of voltage-controlled oscillator based analog-to-digital converter," *IEEE Trans. Circuits Syst. I Regul. Pap.*, vol. 57, no. 1, pp. 18–30, 2010.

[7] J. Van Rethy and G. Gielen, "An energy-efficient capacitance-controlled oscillator-based sensor interface for MEMS sensors," *Proc. 2013 IEEE Asian Solid-State Circuits Conf. A-SSCC 2013*, pp. 405–408, 2013.

[8] X. Wang, J. Zhao, Y. Zhao, G. M. Xia, A. P. Qiu, Y. Su, and Y. P. Xu, "A 0.4 µg Bias Instability and 1.2 µg/√Hz Noise Floor MEMS Silicon Oscillating Accelerometer With CMOS Readout Circuit," *IEEE J. Solid-State Circuits*, vol. 52, no. 2, pp. 472–482, Feb. 2017.

[9] Kenneth Edward Wojciechowski, "Electronics for Resonant Sensors," UNIVERSITY of CALIFORNIA, berkeley, USA, 2015.

[10] I. Galton and G. Zimmerman, "Combined Rf Phase Extraction and Digitization," *1993 Ieee Int. Symp. Circuits Syst. Proceedings, Vols 1–4 (Iscas 93)*, pp. 1104–1107, 1993.

[11] F. Colodro and A. Torralba, "Frequency-to-digital conversion based on sampled phase-locked loop with third-order noise shaping," *Electron. Lett.*, vol. 47, no. 19, p. 1069, 2011.

[12] F. Colodro and A. Torralba, "Frequency-to-digital conversion based on a sampled Phase-Locked Loop," *Microelectronics J.*, vol. 44, no. 10, pp. 880–887, Oct. 2013.

[13] Jian Zhao, Xi Wang, Yang Zhao, Guo Ming Xia, An Ping Qiu, Yan Su, and Yong Ping Xu, "A 0.23 µg bias instability and 1.6 µg/Hz$^{1/2}$ resolution silicon oscillating accelerometer with build-in $\Sigma - \Delta$ frequency-to-digital converter," in 2016 *IEEE Symposium on VLSI Circuits (VLSI-Circuits)*, 2016, pp. 1–2.

[14] J.-J. E. Slotine and W. Li, *Applied Nonlinear Control*, vol. 62, no. 7. 1991.

[15] J. Zhao, Y. Zhao, X. Wang, G. Xia, A. Qiu, Y. Su, and Y. P. Xu, "A System Decomposition Model for Phase Noise in Silicon Oscillating Accelerometers," *IEEE Sens. J.*, vol. 16, no. 13, pp. 5259–5269, 2016.

[16] L. Aaltonen and K. A. I. Halonen, "An analog drive loop for a capacitive MEMS gyroscope," *Analog Integr. Circuits Signal Process.*, vol. 63, no. 3, pp. 465–476, 2010.

[17] X. Wang, J. Zhao, Y. Zhao, G. M. Xia, A. P. Qiu, Y. Su, and Y. P. Xu, "A 0.4 µg Bias Instability and 1.2 µg/√Hz Noise Floor MEMS Silicon Oscillating Accelerometer With CMOS Readout Circuit," *IEEE J. Solid-State Circuits*, pp. 1–11, 2016.

[18] R. Hopkins, J. Miola, and R. Setterlund, "The silicon oscillating accelerometer: A high-performance MEMS accelerometer for precision navigation and strategic guidance applications," *Proc. Annu. Meet. Inst. Navig.*, pp. 1043–1052, 2005.

[19] M. Yücetaş, M. Pulkkinen, A. Kalanti, J. Salomaa, L. Aaltonen, and K. Halonen, "A high-resolution accelerometer with electrostatic damping and improved supply sensitivity," *IEEE J. Solid-State Circuits,* vol. 47, no. 7, pp. 1721–1730, 2012.

[20] P. Ullah, V. Ragot, P. Zwahlen, and F. Rudolf, "A New High Performance Sigma-Delta MEMS Accelerometer for Inertial Navigation," in *Inertial Sensors and Systems Symposium (ISS), 2015 DGON,* pp. 1–13, 2015.

[21] H. Xu, X. Liu, and L. Yin, "A closed-loop $\Sigma\Delta$ interface for a high-Q micromechanical capacitive accelerometer with 200 ng/$\sqrt{Hz}$ input noise density," *IEEE J. Solid-State Circuits,* vol. 50, no. 9, pp. 2101–2112, 2015.

[22] B. V. Amini and F. Ayazi, "Micro-gravity capacitive silicon-on-insulator accelerometers," *J. Micromechanics Microengineering,* vol. 15, no. 11, pp. 2113–2120, 2005.

[23] P. Zwahlen, Y. Dong, A. M. Nguyen, F. Rudolf, J. M. Stauffer, P. Ullah, and V. Ragot, "Breakthrough in high performance inertial navigation grade Sigma-Delta MEMS accelerometer," in *Record – IEEE PLANS, Position Location and Navigation Symposium,* pp. 15–19, 2012.

[24] T. a. Roessig, R. T. Howe, a. P. Pisano, and J. H. Smith, "Surface-micromachined resonant accelerometer," *in Proceedings of International Solid State Sensors and Actuators Conference (Transducers '97),* vol. 2, pp. 1–4, 1997.

[25] A. A. Seshia, M. Palaniapan, T. A. Roessig, R. T. Howe, R. W. Gooch, T. R. Schimert, and S. Montague, "A vacuum packaged surface micro-machined resonant accelerometer," *J. Microelectromechanical Syst.,* vol. 11, no. 6, pp. 784–793, 2002.

[26] A. A. Trusov, S. A. Zotov, B. R. Simon, and A. M. Shkel, "Silicon accelerometer with differential Frequency Modulation and continuous self-calibration," in *Proceedings of the IEEE International Conference on Micro Electro Mechanical Systems (MEMS),* pp. 29–32, 2003.

5

A System-decomposition Model for MEMS Silicon Oscillating Accelerometer

Jian Zhao[1], Yong Ping Xu[2] and Yan Su[3]

[1]Tsinghua University, Beijing, China
[2]Department of Electrical & Computer Engineering, National University of Singapore, Singapore
[3]Nanjing University of Science and Technology, Nanjing, China
E-mail: zhaojianycc@mail.tsinghua.edu.cn

In this chapter, a phase noise model for MEMS Silicon Oscillating Accelerometers (SOA) is described. This model employs a unified approach to decompose the physical system of the SOA into phase and amplitude subsystems to predict the phase noise of the MEMS oscillators and the performance of the MEMS SOAs. It reveals the impacts of various noises, in both MEMS transducer and readout circuits, on the SOA performance. The soundness of this model is validated against time-domain numerical simulations.

5.1 Introduction

As input acceleration to a MEMS SOA is frequency modulated, any noise sources that cause fluctuations in oscillation frequency will deteriorate the performance. Thus, phase noise analysis plays an important role in the MEMS SOA design. It dictates several key regions in the output noise spectrum of the SOA, thereby determines the key specifications, such as bias-instability, resolution, and bandwidth, of the SOA.

In general, the phase noise spectrum can be expressed by a polynomial [1]:

$$S_\phi(f) = \sum_n h_n f^n \tag{5.1}$$

where h_n is a coefficient and n is the order of each term. In an electronic oscillator, the phase noise usually consists of four segments with different n, namely, random walk of frequency ($n = -4$), flicker frequency noise ($n = -3$), white frequency noise ($n = -2$) and white phase noise ($n = 0$), as shown in Figure 5.1(a). As the instantaneous frequency is a derivation of the phase, with a given phase noise spectrum, the corresponding frequency noise spectrum can be obtained with Equation (5.2) and shown in Figure 5.1(b),

$$S_\psi(f) = f^2 S_\phi(f) = \sum_n h_n f^{n+2} \tag{5.2}$$

Several phase noise models have been reported to predict the phase noise in electronic oscillators. Among them, the Linear Time Invariant (LTI) model

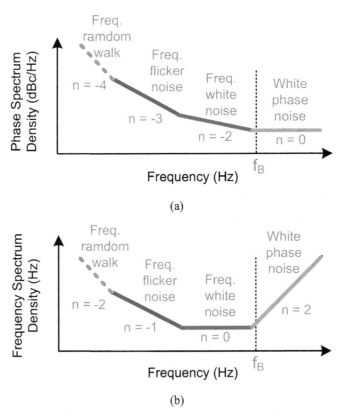

(a)

(b)

Figure 5.1   (a) Typical phase noise spectrum and (b) its equivalent frequency noise spectrum.

in [2, 3] is simple and easy to understand, but fails to reveal the noise mechanism in low frequency region, such as *1/f³* phase noise. Later the Linear Time Variant (LTV) model is proposed to deal with such drawbacks [4]. The LTV model reveals that the phase noise in the *1/f³* region is induced by the asymmetric state-space trajectory and achieves better accuracy in phase noise prediction.

Although the LTV phase noise model can be applied to MEMS oscillator, it alone cannot accurately predict the phase noise. In LTV model, it assumes that the amplitude noise is suppressed by the amplitude-limiting mechanism in the oscillator and doesn't introduce any phase noise. However, this is not true in the MEMS oscillator. As the stiffness of MEMS resonator is nonlinear, the motion equation of MEMS resonator includes a nonlinear term proportional to x^3:

$$m_0 \frac{d^2 x}{dt^2} + c_0 \frac{dx}{dt} + k_0 x + k_2 x^3 = F \tag{5.3}$$

$$F_{elastic} = k_0 x + k_2 x^3 \tag{5.4}$$

where x is the displacement of the resonator beam with an equivalent mass of m_0, c_0 is the damping coefficient, k_0 and k_2 are the linear and cubic nonlinear stiffness, respectively, and F and $F_{elastic}$ are the excitation force and elastic force, respectively. With the nonlinear stiffness k_2, the equivalent spring gets "harder" or "softer" when resonator displacement amplitude changes, which subsequently changes the resonant frequency of the MEMS resonator. This phenomenon is widely known as amplitude-stiffening (A-S) effect. Due to this effect, in MEMS SOAs, the noise in the signal amplitude that drives the MEMS resonator will couple to the oscillation frequency of the MEMS oscillator and generate additional phase noise.

In order to take the A-S effect into account, several modified models have been proposed [5–7]. Although these models improve the phase noise prediction in different ways, there is not a unified approach to deal with practical noise sources in MEMS oscillators and SOAs.

This chapter introduces a system decomposition phase noise model that adopts a unified approach to deal with phase noise analysis in the MEMS SOA. The model identifies the key physical noise sources in MEMS SOA systems and classifies them into either additive or multiplicative noises [8]. According to the impacts of these noises on the performance, the SOA system is decomposed into two paths, namely, the phase modulation (PM) path and amplitude modulation (AM) path. With the system decomposition, the phase/frequency noise analysis can be simplified. Since this model links the performance to the practical noise sources, it provides better guidelines for MEMS oscillator or SOA designs.

The rest of this chapter is organized as follows. Section 5.2 introduces the concept of the MEMS SOA. Section 5.3 introduces physical noise sources in the SOA system, including both mechanical and electronic noises. Section 5.4 classifies these noises into different categories. Section 5.5 describes the system decomposition phase noise model through a bottom-up approach, in which a modulation matrix is introduced. Section 5.6 explains how to apply this model to a practical example with a step-by-step guide. Section 5.7 verifies the system decomposition phase noise model with numerical simulations, and Section 5.8 concludes the chapter.

5.2 Silicon Oscillating Accelerometer

A conceptual schematic of the MEMS SOA is shown in Figure 5.2. When the proof mass is subject to acceleration force along the y axis (which is the sensing direction), it loads two resonators on each side of the proof mass and change their stiffness. As a result, the natural frequencies of the two resonators will change in opposite directions. Two MEMS oscillators are formed with each resonator, respectively, to track the changes of resonant frequencies. In practice, the resonant frequency is not only affected by the input acceleration, but also the noises in the system, which deteriorates the performance.

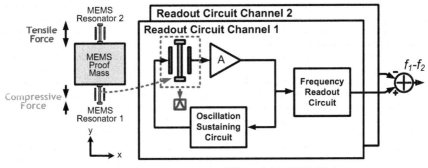

Figure 5.2 Schematic of a differential MEMS SOA.

As mentioned before, the nonlinear stiffness causes the resonant frequency to change with the amplitude. The relationship is given by [9]:

$$\omega_0' = \omega_0 \left(1 + \frac{3k_2}{8k_0} x_0^2 \right) = \omega_0 \left(1 + \lambda x_0^2 \right) \tag{5.5}$$

where x_0 is the displacement amplitude, ω_0' and ω_0 are the resonant frequencies with and without nonlinear stiffness, respectively, and k_0 and k_2 are the linear and cubic nonlinear spring constants, respectively. Equation (5.5) implies that with the nonlinear stiffness, any fluctuation in the oscillation amplitude can cause an oscillation frequency fluctuation.

5.3 Noise Sources

5.3.1 Mechanical Noises

MEMS moving structures have tiny size and ultra-light weight, especially for the resonators in the MEMS SOA, whose dimension is usually in the micron level, and the weight is only several micrograms. This makes the MEMS structure susceptible to molecular agitations and hence generates so-call mechanical noise. As stated in [10], any mechanical system can be analyzed for mechanical thermal noise by adding a force generator alongside each damper. And the force spectral density is given by:

$$F_{thermal} = \sqrt{4k_B TR} \tag{5.6}$$

where k_B is the Boltzmann's constant, T is absolute temperature, R represents the equivalent motional resistance of a mechanical system. The mechanical thermal noise is typically a white noise source. For the resonator in Figure 5.2, the molecular agitation along the x axis will generate a noise n_m in the driving force for the resonator, while the noise force on the proof mass in y axis will load the resonator and induce a noise strain in resonant beam, which is denoted by n_e. This noise strain is directly superimposed on the input acceleration and caused the acceleration noise.

5.3.2 Electronic Noises

The electronic noises in a MOSFET can be modeled in Figure 5.3. The overall noise can be represented by two voltage noise sources on the gate, which are the white thermal noise source, V_n, and the flicker noise, V_{fn}, respectively.

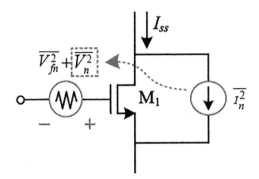

Figure 5.3   Input-referred noise of a MOSFET.

The noise spectrum density of these two noise sources are, respectively, given by:

$$\overline{V_n^2} = \frac{4k_B T \gamma}{g_m} \tag{5.7}$$

$$\overline{V_{fn}^2} = \frac{K}{C_{ox} W L} \frac{1}{f} \tag{5.8}$$

where γ is the constant related to the operation regions of the MOSFET, g_m is the transconductance, C_{ox} denotes the gate capacitance per unit area, W and L are the channel width and length of MOSFET, respectively, and K is a constant related to the process.

Since the above-mentioned mechanical and electronic noises are either white or flicker noises or contain both, it is reasonable to model each noise with a flicker noise and a white noise component, and the noise spectrum density (PSD) can be expressed by:

$$S_i(\omega) = \alpha_i + \frac{\beta_i}{\omega} \tag{5.9}$$

where α_i and β_i are the PSDs of the white and flicker noise at node i, respectively. α_i or β_i can be set to zero if one of the components is absent.

5.4 Noise Classification

The noise sources in the oscillator are classified into three categories based their characteristics, namely, additive noise (n_{add}), multiplicative noise (n_{mul}) and stiffness modulation noise (n_{stiff}). Figure 5.4 illustrates these three noise categories.

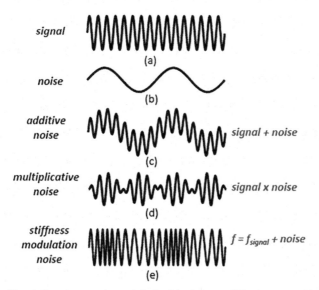

Figure 5.4 The influences to the original signal from different categories of noises. (a) original oscillation signal; (b) sine wave noise; (c) oscillation signal contaminated by additive noise; (d) oscillation signal contaminated by multiplicative noise; and (e) oscillation signal contaminated by stiffness modulation noise.

5.4.1 Additive and Multiplicative Noises

The additive noise, as its name implies, is directly added to the original signal. As shown in Figure 5.4(c), the additive noise is superimposed on the original signal. Multiplicative noise is considered as a noise that causes gain fluctuation. As shown in Figure 5.4(d), a signal is disturbed by the multiplicative noise, whose envelope becomes noisy.

5.4.2 Stiffness Modulation Noise

The natural frequency of the MEMS resonator is $\sqrt{k_{eff}/m_0}$, where k_{eff} and m_0 are nominal stiffness and resonator beam mass, respectively. The variation of the resonator beam mass can be neglected. However, the stiffness k_{eff}, is sensitive to the force applied to the resonator beam and the displacement of the beam, and thus perturbs the natural frequency or the oscillation frequency. Figure 5.4(e) shows the signal disturbed by stiffness modulation noise, which effectively modulate the oscillation frequency.

When the nonlinear stiffness term, k_2, is considered, the stiffness k_{eff} is subject to A-S effect and electrostatic stiffness effects, respectively.

From Equation (5.4), the sensitivity induced by A-S effect from x_0 to k_{eff} can be derived as follows:

$$\frac{\partial k_{eff}}{\partial x} = 2k_2 x_0 \tag{5.10}$$

On the other hand, the stiffness is also affected by the time-variant strain in the resonator beam, which is perturbed by the fluctuations in the bias voltage and input acceleration. The relationship between bias voltage, V_b, and effective stiffness is given by [11]:

$$\frac{\partial k_{eff}}{\partial V_b} = -2\varepsilon_0 k_{vp} V_b \tag{5.11}$$

where ε_0 is the permittivity of vacuum and k_{vp} is a constant dependent on the dimension of variable-gap capacitors in driving and sensing electrodes. From (5.11), a fluctuation in the bias voltage will impact on the oscillation frequency.

The mechanical noise along y axis is equivalent to a random input-referred acceleration noise, the sensitivity of effective stiffness to the strain fluctuation has been derived in [12]:

$$\frac{\partial k_{eff}}{\partial e} = \frac{EA_0}{L_0} \int_0^1 \left(\frac{d\phi_i}{d\varepsilon}\right)^2 d\varepsilon \tag{5.12}$$

where E is the elastic modulus of the resonant beam, A_0 and L_0 are the cross-section area and length of the resonant beam, $\phi_i(\varepsilon)$ is the ith order mode shape function of the resonant beam, and ε is defined as y/L_0, to normalize the mode shape function. The mechanical noise induced effective stiffness fluctuation will eventually be modulated to the oscillation frequency and deteriorate the performance of the accelerometer.

5.4.3 Noise Classification Examples

Figure 5.5 is a block diagram of a typical MEMS oscillator, which consists of a MEMS resonator, an oscillation sustaining circuit, and an Auto-Amplitude Control (AAC) circuit that regulates the displacement amplitude of the MEMS resonator. The major noise sources in the oscillator system are also included in the block diagram and classified into three categories as defined early. Another topology of MEMS oscillator based on a phase locked loop (PLL) in [13, 14] is shown in Figure 5.6, with all noise sources.

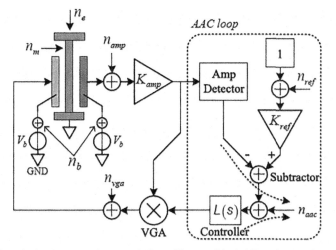

Figure 5.5 Block diagram of an MEMS oscillator with AAC and all noise sources [8]. (Courtesy of IEEE).

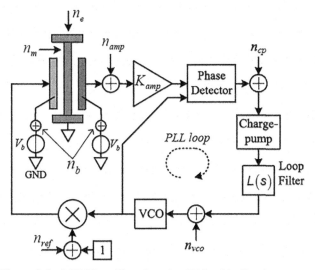

Figure 5.6 MEMS oscillator based on PLL with all noise sources.

Table 5.1 lists all the noises in Figures 5.5 and 5.6, together with their classifications. Note that n_b, not only changes the loop gain, but also directly changes the stiffness through electrostatic spring softening effect, and therefore, it belongs to two categories.

Table 5.1 Noise classification

Noise	Category	Remarks
n_e	stiffness modulation	Strain noise from proof mass
n_m	additive	Mechanical noise along drive direction
n_b	multiplicative, stiffness modulation	Noise of bias voltage on MEMS electrodes
n_{amp}	additive	Input-referred noise of front-end amplifier
n_{vga}	additive	Output noise of VGA in AAC-based system
n_{ref}	multiplicative	Noise in reference voltage for amplitude control
n_{aac}	additive	Lumped additive noise in AAC circuit
n_{cp}	additive	Input noise of charge pump in PLL based system
n_{vco}	additive	Output noise of VCO in PLL-based system

5.5 System Decomposition Model

In practice, various noise sources in the MEMS SOA system, such as additive noises and gain fluctuations, impact on the oscillation frequency differently. If these two mechanisms can be separated or decomposed in the analysis, the order of system can be reduced and the overall phase noise can be analyzed under linear condition without compromising the accuracy of the prediction. In the following subsections, the system decomposition model is introduced and is first applied to a damped MEMS resonator and then to the entire MEMS SOA system, including the readout circuit.

5.5.1 Time-domain Decomposition for Damped MEMS Resonator

In ideal case, the differential motion equation of the MEMS resonator x axis (Referring to Figure 5.2) is

$$m_0 \frac{\mathrm{d}^2 x}{\mathrm{d}t^2} + c_0 \frac{\mathrm{d}x}{\mathrm{d}t} + k_0 x = F \tag{5.13}$$

where x is the displacement of the resonator beam with a lumped mass of m_0, c_0 is the damping coefficient, k_0 is the linear and cubic nonlinear stiffness and F *is* the excitation force. When the resonator is subjected to an ideal sinusoidal driven force with a frequency of ω_0, whose complex expression is given by:

$$F(t) = A e^{j\omega_0 t} \tag{5.14}$$

the displacement response of resonant beam can be obtained,

$$d(t) = \frac{AQ}{m_0\omega_0^2}e^{j\left(\omega_0 t - \frac{\pi}{2}\right)} = K_m A e^{j\left(\omega_0 t - \frac{\pi}{2}\right)} \tag{5.15}$$

where Q and K_m are the quality factor and the gain of the resonator at w_0, respectively. Equation (5.15) can be regarded as the basic response, and the effect of noise and nonlinearity such as chaotic behavior in the Duffing resonators [15] or the electrical saturations [16] can be superposed on the basic response.

Now let us consider an open-loop driving situation with two noise sources at input, as shown in Figure 5.7. n_{add} represents an additive noise source that has zero mean value, while n_{mul} a multiplicative noise source with an unit mean value. First, we only take n_{add} into consideration as an example. If n_{add} is a unit impulse function at moment τ. The response h_{add} can be obtained, and for MEMS resonator with a high Q factor, the response can be approximately given by [17]:

$$h_{add}(t) \approx K_m u(t-\tau) e^{-\frac{\omega_0}{2Q}(t-\tau)+j\left[\omega_0(t-\tau)-\frac{\pi}{2}\right]} \tag{5.16}$$

This equation can be understood as a sinusoidal signal with exponentially decreasing amplitude. When the input of resonator contains both sinusoidal driving force and impulse function, n_{add}, the total response is a superposition of $d(t)$ and $h_{add}(t)$. At a specific moment, the composite response can be depicted in Figure 5.8(a). Then, let us closely look at Figure 5.8(a), compared to the ideal response, the displacement vector in the complex plane changes both its phase and amplitude under the effect of a unit impulse noise, which can be extracted as $h_{add}^{\langle p \rangle}$ and $h_{add}^{\langle a \rangle}$, respectively.

$$h_{add}^{\langle p \rangle}(\tau, t) = h_r^{\langle p \rangle}(\tau, t) \frac{\sin(\omega_0 \tau)}{A} \tag{5.17}$$

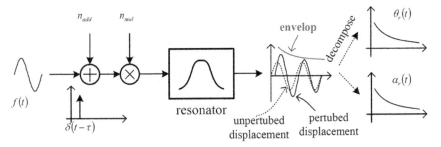

Figure 5.7 Amplitude and phase responses under a perturbed sinusoidal input [8]. (Courtesy of IEEE).

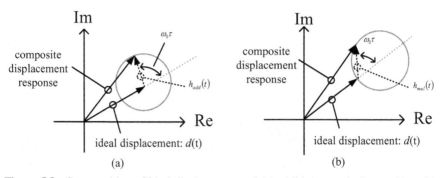

Figure 5.8   Superposition of ideal displacement and (a) additive perturbation or (b) multiplicative perturbation [8]. (Courtesy of IEEE).

$$h_{add}^{\langle a \rangle}(\tau, t) = h_r^{\langle a \rangle}(\tau, t) \cos(\omega_0 \tau) \tag{5.18}$$

It can be seen from Equations (5.17) and (5.18), both phase and amplitude responses are dependent on the impulse moment τ. Moreover, the phase response is inversely proportional to driving amplitude, A. The remnant terms $h_r^{\langle p \rangle}$ and $h_r^{\langle a \rangle}$, are given by, respectively,

$$h_r^{\langle p \rangle}(t) = e^{-\frac{\omega_0}{2Q}(t)} u(t) \tag{5.19}$$

$$h_r^{\langle a \rangle}(t) = K_m e^{-\frac{\omega_0}{2Q}(t)} u(t) \tag{5.20}$$

where $h_r^{\langle p \rangle}$ and $h_r^{\langle a \rangle}$ are exponential decay functions that can be regarded as two projections of h_{add} in the directions of phase and amplitude of carrier signal, respectively.

Next, let us take n_{mul} into consideration, as mentioned before, we assumed that n_{mul} is a unit impulse. As shown in Figure 5.7, this noise modulates the force $f(t)$ and then drives the resonator. The driving force induced by n_{mul} is given by:

$$\delta_{mul}(t - \tau) = A \cos(\omega_0 t) \delta(t - \tau) \tag{5.21}$$

Similar to the analysis of n_{add}, for n_{mul}, the phase and amplitude impulse responses, $h_{mul}^{\langle p \rangle}$ and $h_{mul}^{\langle a \rangle}$, can be obtained below:

$$h_{mul}^{\langle p \rangle}(\tau, t) = h_r^{\langle p \rangle} \frac{\sin(2\omega_0 \tau)}{2} \tag{5.22}$$

$$h_{mul}^{\langle a \rangle}(\tau, t) = h_r^{\langle a \rangle} \frac{A}{2}[1 - \cos(2\omega_0 \tau)] \tag{5.23}$$

The overall response for the multiplicative impulse is shown in Figure 5.8(b). The difference between the additive and multiplicative perturbations is that the length of the additive perturbation vector is independent of τ, while that of the multiplicative perturbation vector is a function of τ.

5.5.2 Frequency-domain Decomposition for Damped MEMS Resonator

The analysis can be further extended to a more general case where the input noises are arbitrary signals, n_{add} and n_{mul}. Take n_{add} as an example, the phase response can be calculated using superposition integral, which is given below [18]:

$$\theta_{add}(t) = \int_{-\infty}^{\infty} h_{add}^{\langle p \rangle}(t,\tau)\, n_{add}(\tau)\, d\tau = h_r^{\langle p \rangle}(t) * \frac{\sin(\omega_0 t)\, n_{add}(t)}{A}$$

(5.24)

where $*$ is the convolution operator. Similarly, the remnant phase θ and amplitude x responses of additive and multiplicative noise can be expressed as follows

$$\theta_{mul}(t) = h_r^{\langle p \rangle}(t) * \frac{\sin(2\omega_0 t)\, n_{mul}(t)}{2}$$

(5.25)

$$x_{add}(t) = h_r^{\langle a \rangle}(t) * \cos(\omega_0 t)\, n_{add}(t)$$

(5.26)

$$x_{mul}(t) = h_r^{\langle a \rangle}(t) * \frac{A\left[1 + \cos(2\omega_0 t)\right] n_{mul}(t)}{2}$$

(5.27)

Then, the total phase and amplitude responses can be derived as:

$$\theta_r(t) = h_r^{\langle p \rangle}(t) * \left[\frac{\sin(\omega_0 t)}{A} n_{add}(t) + \frac{\sin(2\omega_0 t)}{2} n_{mul}(t) \right]$$
$$= h_r^{\langle p \rangle}(t) * n^{\langle p \rangle}(t)$$

(5.28)

$$x_r(t) = h_r^{\langle a \rangle}(t) * \left[\cos(\omega_0 t)\, n_{add}(t) + \frac{A\left[1 + \cos(2\omega_0 t)\right]}{2} n_{mul}(t) \right]$$
$$= h_r^{\langle a \rangle}(t) * n^{\langle a \rangle}(t)$$

(5.29)

Note that the first term of all phase and amplitude responses can be identified as an impulse response of a first-order system. The corresponding transfer functions are derived in Equations (5.30) and (5.31) through Laplace transform, which can also be defined as the PM and AM transfer functions of the

MEMS resonator, respectively,

$$H_r^{\langle p \rangle}(s) = \frac{\omega_0/(2Q)}{s + \omega_0/(2Q)} = \frac{\omega_c}{s + \omega_c} \tag{5.30}$$

$$H_r^{\langle a \rangle}(s) = K_m \frac{\omega_c}{s + \omega_c} \tag{5.31}$$

where ω_c is the cut-off frequency of decomposed transfer function. The second term can be regarded as the pre-modulation of the input noise before applying to the decomposed system. Note that the PM transfer function has a unit DC gain, while the gain of AM transfer function is equal to K_m. Based on the above analyses, the original high-order transfer function of MEMS resonator has been replaced by two lower order subsystems, whose outputs are directly phase and amplitude noises.

5.5.3 Modulation Matrix

To facilitate the analysis, from Equation (5.28) and (5.29), it is convenient to introduce a modulation matrix, which can convert the physical noises n_{add} and n_{mul} to the decomposed noises $n^{\langle p \rangle}(t)$ and $n^{\langle a \rangle}(t)$, respectively. The modulation matrix is a function of t and denoted by M,

$$\begin{bmatrix} n^{\langle p \rangle}(\tau) & n^{\langle a \rangle}(\tau) \end{bmatrix} = M^T(\tau) \begin{bmatrix} n_{add}(\tau) & n_{mul}(\tau) \end{bmatrix}^T \tag{5.32}$$

where

$$M(\tau) = \begin{bmatrix} A^{-1}\sin(\omega_0\tau) & \cos(\omega_0\tau) \\ 0.5\sin(2\omega_0\tau) & 0.5A\left[1 + \cos(2\omega_0\tau)\right] \end{bmatrix} \tag{5.33}$$

The output phase and amplitude noises can be obtained by applying PM and AM transfer functions to the corresponding noise inputs, respectively, that is,

$$\begin{bmatrix} \theta_r(t) \\ x_r(t) \end{bmatrix} = \int_{-\infty}^{\tau} \begin{bmatrix} h_r^{\langle p \rangle} \\ & h_r^{\langle a \rangle} \end{bmatrix} \begin{bmatrix} n^{\langle p \rangle}(\tau) & n^{\langle a \rangle}(\tau) \end{bmatrix}^T d\tau \tag{5.34}$$

Based on Equations (5.32) to (5.34), the decomposed resonator can be modeled by the block diagram in Figure 5.9, and the phase and amplitude fluctuations induced by the additive and multiplicative noises under open-loop driving condition can be easily obtained.

5.5.4 Decomposition of a Practical MEMS Oscillation System

Now, let us apply the above analysis to a practical MEMS oscillator system. For simplicity, the bias voltage V_b and front-end gain K_{amp} of MEMS oscillator in Figure 5.5 is assumed to be 1. By applying the decomposed resonator

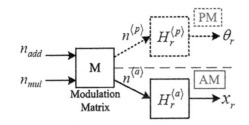

Figure 5.9 Equivalent block diagram of the decomposed MEMS resonator [8]. (Courtesy of IEEE).

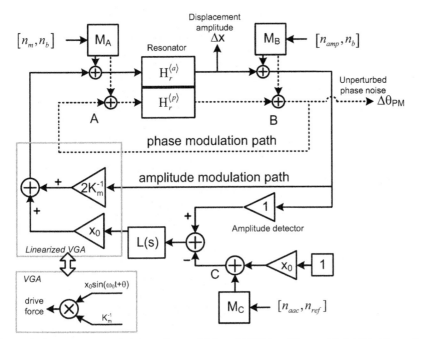

Figure 5.10 Decomposed phase noise model for a MEMS oscillator with AAC in Figure 5.5 [8]. (Courtesy of IEEE).

model, the overall MEMS oscillator system can be decomposed into AM and PM paths and shown in Figure 5.10. Based on a pre-set reference, the steady amplitude of each node can be obtained and the corresponding modulation matrix can also be calculated. Then the phase and amplitude fluctuations can be estimated according to Figure 5.9.

Figure 5.10 shows the decomposed system block diagram for the MEMS oscillator with AAC, such as the one in Figure 5.5. Noted that the system only processes amplitude information between the amplitude detector and the VGA. As a result, the first column of the modulation matrix in this region is zero. For example, the modulation matrix at node C in Figure 5.10 is given by

$$M_C = \begin{bmatrix} 0 & 1 \\ 0 & A_C \end{bmatrix} \tag{5.35}$$

Similarly, for the PLL-based oscillator in Figure 5.6, it can be decomposed as shown in Figure 5.11. As phase information is demodulated by phase detector, so the modulation matrix at node C and D only have phase responses, therefore, the second column is zero. For example, the modulation matrix at node D is given by:

$$M_D = \begin{bmatrix} 1 & 0 \\ A_D & 0 \end{bmatrix} \tag{5.36}$$

5.5.5 Phase Noise Modeling of Entire MEMS SOA Encompassing Nonlinearities

Based on the decomposed linear models in Figures 5.10 and 5.11, the proposed model can already predict the phase noise and amplitude noise of an oscillation system. However, as nonlinearities exist in practical systems, there are always some couplings from the amplitude fluctuations or other factors to the oscillation frequency as illustrated in Section 5.4.2.

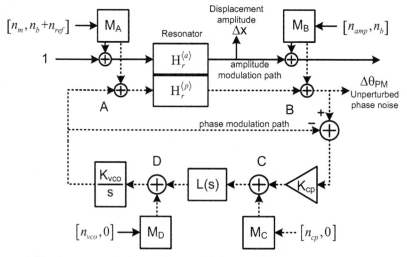

Figure 5.11   Decomposed phase noise model for a MEMS oscillator based on PLL in Figure 5.8.

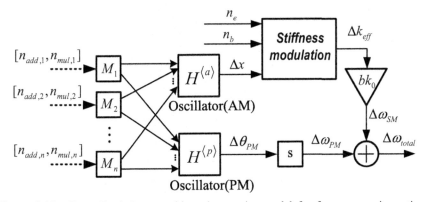

Figure 5.12 Generalized decomposition phase noise model for frequency noise estimation [8]. (Courtesy of IEEE).

Take the nonlinearities into consideration, a generalized multi-input MEMS SOA phase noise model can be established for frequency noise analysis and is shown in Figure 5.12. In this model, the physical noise sources are described as a noise vector that represents the additive and multiplicative noise at a specific node. Then, each noise (additive or multiplicative) can be decomposed into AM and PM noise through the corresponding modulation matrix. After the decomposition, the noise sources are transformed into amplitude fluctuation (Δx) and phase fluctuation ($\Delta\theta_{PM}$) through the decomposed sub-systems. Finally, the nonlinearity-induced frequency fluctuations are calculated based on the stiffness modulation mechanisms and their corresponding inputs. Meanwhile, the phase noise is directly translated to the frequency fluctuation through differentiation. The overall frequency fluctuation (frequency noise) is the sum of the two components.

5.6 Noise Estimation with System Decomposition Phase Noise Model

This section will illustrate step by step how to predict the phase noise of MEMS SOA using the system decomposition phase noise model. An AAC-based MEMS oscillator is used as an example and shown in Figure 5.13. Four typical noise sources, n_m at node A, n_{amp} and n_{bs} at node B, and n_{ref} at node C, are included in the system.

The key parameters of the system are listed in Table 5.2. In order to shorten the transient simulation time, the resonant frequency of the MEMS resonator is set to be 10 rad/s. All the gain blocks in the loop are set to 1

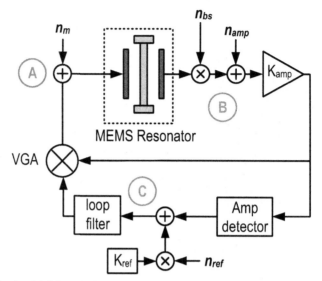

Figure 5.13   An AAC-based MEMS SOA with four typical noise sources [8].
(Courtesy of IEEE).

Table 5.2   Model Parameters for Simulations in SIMULINK

Parameter's name	Symbol	Unit	Value
Beam mass	m_0	kg	1
zero-order stiffness	k_0	N/m	100
2nd-order stiffness	k_2	N/m^3	1000
Drive force/drive voltage	K_{bd}	N/V	1
Sensing current/resonator velocity	K_{bs}	A/m/sec	1
Damping coefficient	c_0	N/m/sec	0.1
Gain of amplifier	K_{amp}	ohm	1
Gain of reference	K_{ref}	1	0.1
Sensitivity from bias voltage to k_{eff}	K_s	N/m/V	25
Loop filter in AAC	$H_c(s)$	1	$\frac{s+0.01}{s\left(s^2+2s+1\right)}$

for simplicity, including the mechanical and electrical interfaces (K_{bd}, K_{bs}), and the front amplifier gain (K_{amp}). It should be noted that the parameters in Table 5.2 are not realistic values and chosen for illustration purpose only.

Step 1 – Preparing the decomposed noise sources: As stated before, the modulation matrix can convert the physical noise sources to decomposed noise sources. Take node B as an example, where there are two noise sources, i.e., n_{amp} and n_{bs}. From Table 5.2, the oscillating amplitude at node B can

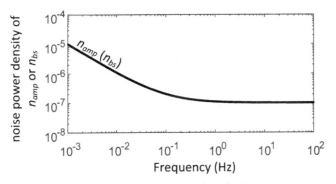

Figure 5.14 Power spectrum density of noise sources.

be calculated as unity. Substituting the amplitude into Equation (5.33), the modulation matrix at node B is given by:

$$M_B\left(\tau\right) = \begin{bmatrix} \sin\left(10\tau\right) & \cos\left(10\tau\right) \\ 0.5\sin\left(20\tau\right) & 0.5\left[1 + \cos\left(20\tau\right)\right] \end{bmatrix} \qquad (5.37)$$

Assume that all the noise sources in Figure 5.13 have the same power spectrum density as shown in Figure 5.14, whose expressions are given by:

$$S_{unit}\left(f\right) = 1e^{-8}\frac{1}{f} + 1e^{-7} \qquad (5.38)$$

Based on Equation (5.32), the physical noise sources, n_{amp} (additive) and n_{bs} (multiplicative), are converted into decomposed noise sources, $n_B^{\langle p\rangle}$ and $n_B^{\langle a\rangle}$, which are shown in Figure 5.15.

Step 2 – Decomposing the system transfer functions: As illustrated in Section 5.5.4, the overall system can be decomposed into phase-mode subsystem and amplitude-mode subsystem in Figure 5.10, in which the two subsystems can also be separately depicted in Figure 5.16, respectively. Based on Equations (5.30) and (5.31), the decomposed transfer functions of the MEMS resonator are given by:

$$H_r^{\langle p\rangle} = H_r^{\langle a\rangle} = \frac{0.05}{s + 0.05} \qquad (5.39)$$

Then, the decomposed transfer functions from node B to the corresponding outputs of the phase- or amplitude-mode subsystems can be easily obtained from Figure 5.16 and their magnitude responses are shown in Figure 5.17.

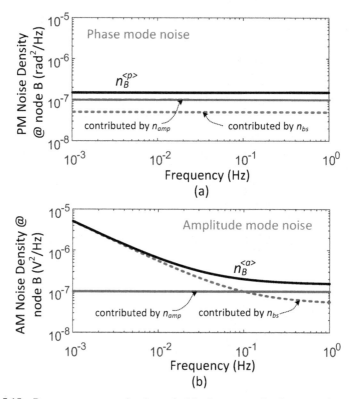

Figure 5.15 Power spectrum density of (a) decomposed phase mode noise and (b) decomposed amplitude mode noise at node B.

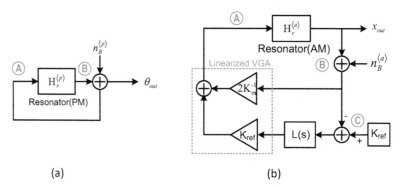

Figure 5.16 (a) Decomposed phase mode subsystem and (b) amplitude mode subsystem [8]. (Courtesy of IEEE).

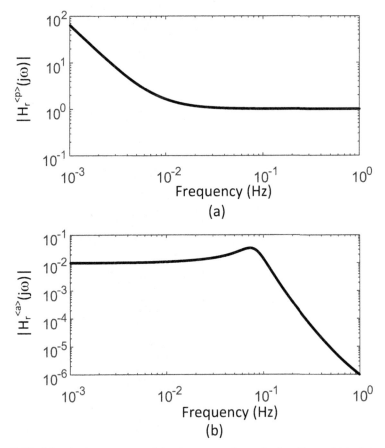

Figure 5.17 Magnitude responses of the decomposed transfer functions of (a) phase mode subsystem and (b) amplitude mode subsystem.

Step 3 – Calculating the outputs of the decomposed phase and amplitude sub-systems: When both decomposed transfer functions and noise sources are known, the output from the decomposed phase and amplitude sub-systems can be calculated, respectively, which are given in Figure 5.18. Note that the outputs from the decomposed phase and amplitude sub-systems are phase and amplitude noise power, respectively.

Step 4 – Converting the phase and amplitude noises into frequency noise: Once the outputs of two sub-systems are known, the overall frequency noise from the oscillator can be calculated. The phase noise can be directly converted to frequency noise by its derivative and the amplitude noise can be

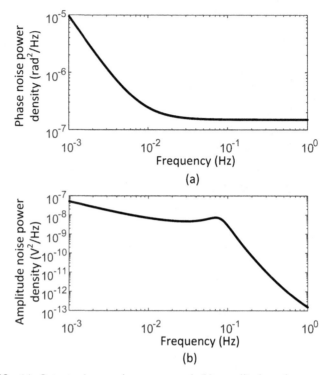

Figure 5.18 (a) Output phase noise power and (b) amplitude noise power from the decomposed phase and amplitude subsystems, respectively.

converted to frequency noise via the A-S effect based on Equation (5.5). The overall frequency noise power is the summation of the two frequency noise powers from the decomposed phase and amplitude sub-systems. Figure 5.19 (a) shows the calculated overall frequency noise due to the two noise sources at node B, and the respective noise components from the phase- and amplitude-mode sub-systems. The contributions from the physical noise sources, n_{amp} and n_{bs}, can also be calculated and are shown in Figure 5.19 (b). Through the above step-by-step illustrations, the results in Figure 5.19 have shown that the proposed model can be used to predict the performance of a MEMS SOA and analyse the contributions to the performance from different mechanisms (phase or amplitude) or different noise sources. This provides very useful guidelines for the MEMS SOA system design, including both MEMS transducer and readout circuit designs.

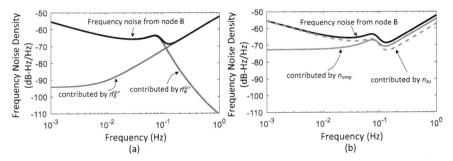

Figure 5.19 Overall frequency noise at the output of the MEMS oscillator due to the noises at node B (a) with noise components from the decomposed phase- and amplitude-mode sub-systems and (b) with two physical noise sources at node B.

Based on the above example, the overall frequency noise contributed by all four noise sources in Figure 5.13 can also be calculated in the similar way, which is shown in Figure 5.20(a). Finally, due to the electrostatic stiffness effect, n_{bs} will contribute another noise component named n_{es} in the final frequency responses, which is shown in Figure 5.20(b). This figure also indicates that the proposed model can be used to reveal the contribution of each noise sources to the overall performance.

The key steps of using system decomposition phase noise model are summarized as follows:

1) Classify the physical noise sources into three categories, i.e., additive, multiplicative and stiffness-modulation noise;
2) Calculate the corresponding modulation matrixes;
3) Convert or decompose the physical noises to the noises for the phase- and amplitude-mode sub-systems using the modulation matrixes;
4) Decompose the system under study into the phase- and amplitude-mode sub-systems;
5) Apply the decomposed noise sources to the corresponding decomposed sub-systems and obtain the respective output amplitude and phase noises;
6) Convert the noises from two sub-systems to frequency noises, respectively, and add the two frequency noise powers to obtain the overall frequency noise power for the system under study;
7) Take nonlinearity (nonlinear stiffness) into consideration to obtain more accurate estimation of the frequency noise.

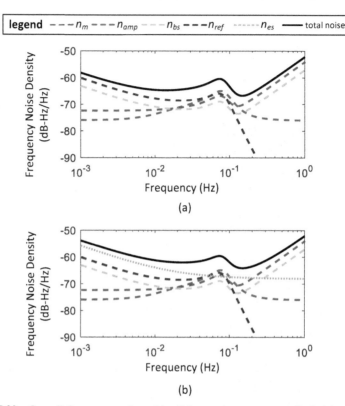

Figure 5.20   Overall frequency noise with all four noise sources applied: (a) without and (b) with the electrostatic stiffness softening.

5.7 Numerical Simulation

To further validate and evaluate the decomposed model, a time-domain numerical model for an AAC-based MEMS oscillator is built in SIMULINK, as depicted in Figure 5.21. The parameters used for simulation are same to those in Table 5.2. In order to shorten the simulation cycle, the quality factor of MEMS resonator is reduced to only 100. This can be calculated from the data in Table 5.2, i.e., quality factor $Q = \frac{\sqrt{m_0 k_0}}{c_0}$.

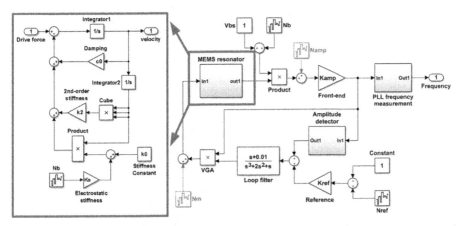

Figure 5.21 SIMULINK time-domain model for AAC-based MEMS SOA [8]. (Courtesy of IEEE).

5.7.1 Performance Prediction

In order to obtain the frequency noise, an ideal PLL is used in SIMULINK to track the oscillation frequency dynamically. The bandwidth of the PLL is set to 2 Hz, which is high enough for observing all slopes in frequency spectrum, as shown in Figure 5.1. All noise sources used in the simulations are assumed to be uncorrelated and consist of both flicker and white noise components. The noise spectrum density of each source is set to be the same, as given in Equation (5.38), whose power spectrum density used in the transient simulation is depicted in Figure 5.22.

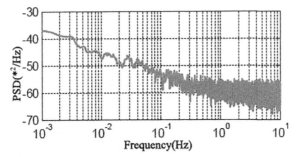

Figure 5.22 Power spectrum density of all noise sources used in the simulation [8]. (Courtesy of IEEE).

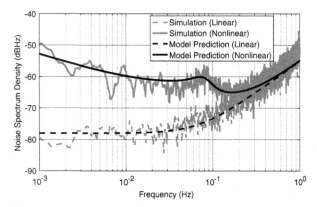

Figure 5.23   Comparison of frequency noise PSDs between the decomposition model and the time-domain simulation for both linear and nonlinear cases [8]. (Courtesy of IEEE).

The first simulation evaluates the performance of the system decomposition model. The noise spectrum densities predicted by the system decomposition model are compared with those from the numerical simulation under both linear and nonlinear cases, which are shown in Figure 5.23. In the linear case, the nonlinearity related parameters k_2 and k_s are set to be zero. Hence, the decomposed model reduces to LTV phase model [4]. In the nonlinear case, the nonlinear stiffness is taken into account and the decomposed model also shows a good agreement with the time domain simulation, especially in the flicker noise dominated low frequency region. Moreover, the Allan variances, which is widely used to evaluate accelerometer performance [19], are also obtained from the above simulations and their comparisons are shown in Figure 5.24.

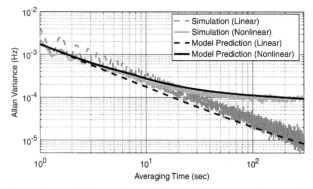

Figure 5.24   Comparison of Allan variances between the decomposition model and the time-domain simulation for both linear and nonlinear cases [8]. (Courtesy of IEEE).

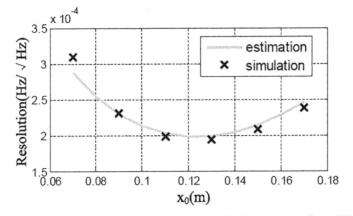

Figure 5.25   In-band noise of MEMS oscillator vs. the displacement amplitude [8]. (Courtesy of IEEE).

5.7.2 The Optimal MEMS Resonator Displacement Amplitude

Another comparison is made to verify the relationship between the in-band frequency noise of the MEMS oscillator under study and the displacement amplitude of the MEMS resonator beam, where for the purpose of demonstration, the frequency noise density within 0.01 to 0.1 Hz bandwidth is considered. The in-band noise is calculated under different displacement amplitudes and plotted in Figure 5.25, together with the time-domain simulation result. A strong agreement is observed between the prediction by the decomposed model and the simulation in SIMULINK. Moreover, if we set the in-band noise as the design target, an optimal displacement can be obtained based on the decomposition model.

5.8 Summary

This chapter introduces a system decomposition phase noise model that can be used to predict the phase and frequency noises of MEMS oscillators and MEMS SOAs. The model is able to reveal how physical noise sources in the SOA contribute to the overall phase or frequency noises and how they impact on SOA performance. Being different from the other phase noise models, the system decomposition phase noise model treats the phase noises caused by the phase and amplitude perturbations separately via decomposing each noise source into these two components. This decomposition facilitates the phase noise analysis and improves the prediction accuracy. The amplitude-stiffness effect that is unique in the MEMS-based oscillator is incorporated in the model through a separate amplitude path. Furthermore, each noise in the system can be modeled by flicker or white noise, or both components, which is close to the real noise sources in most practical systems. The decomposition model has been validated by time-domain simulations in a SIMULINK model. This model can be used to predict the performance of MEMS oscillator and MEMS SOA, and identify the dominant noise sources in the system, which provides useful design guidelines.

Acknowledgement

The author would like to thank Yang Zhao and Xi Wang for the insightful discussions.

References

[1] E. Rubiola, *Phase noise and frequency stability in oscillator.* Cambridge University Press, 2009.

[2] D. B. Leeson, "A Simple Model of Feedback Oscillator Noise Spectrum," *Proc. IEEE,* vol. 54, no. 2, pp. 329–330, 1966.

[3] B. Razavi, "A Study of Phase Noise in CMOS OSC.pdf," *IEEE J. Solid-State Circuits,* vol. 31, no. 3, pp. 331–343, 1996.

[4] A. Hajimiri and T. H. Lee, "A general theory of phase noise in electrical oscillators," *IEEE J. Solid-State Circuits,* vol. 33, no. 2, pp. 179–194, 1998.

[5] L. H. L. He, Y. P. X. Y. P. Xu, and M. Palaniapan, "A State-Space Phase-Noise Model for Nonlinear MEMS Oscillators Employing Automatic Amplitude Control," *IEEE Trans. Circuits Syst. I Regul. Pap.,* vol. 57, no. 1, pp. 189–199, 2010.

[6] R. Shi, F. X. Jia, A. P. Qiu, and Y. Su, "Phase noise analysis of microme-chanical silicon resonant accelerometer," *Sensors Actuators, A Phys.,* vol. 197, pp. 15–24, 2013.

[7] P. Ward and A. Duwel, "Oscillator phase noise: Systematic construc-tion of an analytical model encompassing nonlinearity," *IEEE Trans. Ultrason. Ferroelectr. Freq. Control,* vol. 58, no. 1, pp. 195–205, 2011.

[8] J. Zhao et al., "A System Decomposition Model for Phase Noise in Silicon Oscillating Accelerometers," *IEEE Sens. J.,* vol. 16, no. 13, pp. 5259–5269, 2016.

[9] L. He, Y. P. Xu, and M. Palaniapan, "A CMOS Readout Circuit for SOI Resonant Accelerometer With 4-Bias Stability and 20-Resolution," *Solid-State Circuits, IEEE J.,* vol. 43, no. 6, pp. 1480–1490, 2008.

[10] T. B. Gabrielson, "Mechanical-Thermal Noise in Micromachined Acoustic and Vibration Sensors," *IEEE Trans. Electron Devices,* vol. 40, no. 5, pp. 903–909, 1993.

[11] C. Acar and A. Shkel, *MEMS vibratory gyroscopes: structural approaches to improve robustness. Springer,* 2008.

[12] T. a W. Roessig, "Integrated MEMS Tuning Fork Oscillators for Sensor Applications." pp. 1–147, 2006.

[13] Jian Zhao *et al.,* "A 0.23-μg bias instability and 1.6-μg/Hz$^{1/2}$ resolution silicon oscillating accelerometer with build-in $\Sigma - Delta$ frequency-to-digital converter," in 2016 *IEEE Symposium on VLSI Circuits (VLSI-Circuits),* pp. 1–2, 2016.

[14] J. Zhao et al., "A 0.23-μmg Bias Instability and 1-μmg/$\sqrt{v}$Hz Acceleration Noise Density Silicon Oscillating Accelerometer With Embedded Frequency-to-Digital Converter in PLL," *IEEE J. Solid-State Circuits,* vol. 52, no. 4, pp. 1053–1065, Apr. 2017.

[15] S. L. S. Lee and C. T.-C. Nguyen, "Influence of automatic level control on micromechanical resonator oscillator phase noise," in *IEEE International Frequency Control Symposium and PDA Exhibition Jointly with the 17th European Frequency and Time Forum. Proceedings of the 2003,* vol. 2003, pp. 341–349, 2003.

[16] A. Tocchio, A. Caspani, G. Langfelder, A. Longoni, and E. Lasalandra, "A pierce oscillator for MEMS resonant accelerometer with a novel low-power amplitude limiting technique," 2012 *IEEE Int. Freq. Control Symp. IFCS 2012, Proc.,* pp. 748–753, 2012.

[17] Giorgio Rizzoni and James Kearns, *Principles and Applications of Electrical Engineering,* vol. 3, McGraw Hill, 2015.

[18] T. H. Lee and A. Hajimiri, "Oscillator phase noise: A tutorial," *IEEE J. Solid-State Circuits,* vol. 35, no. 3, pp. 326–335, 2000.

[19] R. Levy and V. Gaudineau, "Phase noise analysis and performance of the Vibrating Beam Accelerometer," in 2010 *IEEE International Frequency Control Symposium, FCS 2010,* pp. 511–514, 2010.

6

Resonant Seismic Sensor

Xudong Zou

State Key Laboratory of Transducer Technology, Institute of Electronics, Chinese Academy of Sciences, Beijing, China
E-mail: xdzou@mail.ie.ac.cn

Micro-electro-mechanical accelerometers (MEMS AXLs) have seen rapid adoption as a low-cost, small size alternative to the traditional accelerometers in various applications. However, the current market is still dominated by MEMS AXLs with mid-to-low resolution, whereas the availability of high-performance devices, e.g. seismic sensors, is very limited. Micro-Electro-Mechanical Resonant Accelerometer (MEMS RXL) is considered to have the potential to address these limitations and achieve high resolution. The MEMS RXL measures the acceleration-induced inertial force on a proof mass using micro-machined vibrating beam gauges. MEMS RXL can endure a very large input acceleration, while simultaneously enabling a high sensitivity measurement, resulting in a high dynamic range. This chapter deals with the design of high-precision MEMS RXLs for seismic acceleration measurement based on state-of-the-art micromachining technologies. The design models as well as noise floor and drift analysis of the MEMS RXL presented herein are intended to instruct the designer to make informed choices in the design of MEMS RXLs for different applications, and to help the designer evaluate, verify, and optimize the performance of MEMS RXLs.

6.1 Introduction

MEMS accelerometers have seen much translational success in a large spectrum of applications ranging from automotive systems, environmental and infrastructure monitoring, user interfaces for mobile and gaming devices, and wearable healthcare. However, a number of emerging applications such as seismology [1–3] require high performance for inertial force measurement. In the area of seismic imaging [3–5], considerably larger-sized geophones are regularly used to conduct seismic surveys. Typically, the signal amplitude of a geophone is linearly proportional to velocity above its resonance frequency, with a roll-off of –40 dB/decade below resonance. MEMS accelerometers, on the other hand, have been demonstrated to provide a relatively flat amplitude and phase frequency response over bandwidths ranging from a few Hz up to 100 Hz. Recent work on the optimization of capacitive [3–5] and optical MEMS accelerometers [6] have also resulted in devices that offer an acceptable resolution, making them potentially attractive for seismic imaging. However, these high-resolution MEMS accelerometers are still limited by their inability to operate over a large dynamic range. Furthermore, the 1/f noise in these devices often limits their low frequency measurement capability with typical measurement frequencies down to a few Hz.

This chapter introduces a new type of seismic sensor based on an MEMS resonant accelerometer (RXL) that offers the advantage of improved electronic noise-limited resolution without sacrificing on the dynamic range and bandwidth of operation. As opposed to the capacitive sensing principles, the resonant accelerometers convert the input acceleration to an inertial force that is sensed by resonant strain gauges coupled to a suspended proof mass. This results in a frequency-modulated output response, which provides an increased immunity to parametric noise and eliminates the trade-off between bandwidth and sensitivity that is inherent in open-loop capacitive or optical accelerometers. The scale factor of the resonant accelerometer remains constant over a large input acceleration range without the need of any additional feedback control, in contrast to the accelerometers based on the displacement measurement principles. The measured acceleration data can be readily demodulated from the quasi-digital output signal by frequency-counting techniques for acceleration frequencies up to several hundred hertz.

This chapter consists of four sections, which cover the analyses and design of sensor structure and readout circuit, the fabrication process, and the noise limited resolution and long-term drift analyses of resonant seismic sensor, respectively.

6.2 Sensor Design

6.2.1 Sensor Topology

The design of the resonant seismic sensor consists of both mechanical and electrical designs, as well as the design of the electro-mechanical transducer.

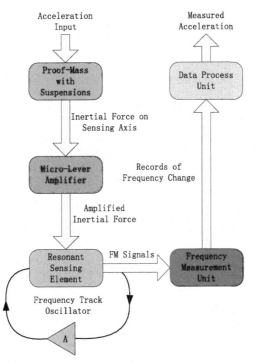

Figure 6.1 Topology of MEMS RXL.

The mechanical part includes resonant sensing elements, inertial force amplifiers, suspension frames and the force feedback actuators. The electrical part includes the DC voltage reference, frequency track oscillator and frequency counter. Figure 6.1 shows the schematic of the MEMS RXL described in this chapter. During the operation, the variance of seismic acceleration will change the inertial force on the proof-mass. The components of the inertial force on a particularly defined axis will be coupled to resonant sensing elements after amplification by micro-levers, but other components of the force will be damped by the suspensions with specific designs. Similar to other resonant MEMS sensors [7–10], the variable inertial force from proof-mass will be converted to the resonant frequency change of the sensing element which will then be transduced to a frequency-modulated (FM) oscillation voltage signal by the frequency tracking oscillator circuit. The FM signal can be recorded by frequency counter or demodulated by PLL-based circuits and then can be back translated to acceleration reading of the RXL.

6.2.2 Mechanical Structure Design

The mechanical structure of the MEMS RXL includes the design of the proof-mass, suspensions, micro-lever force amplifiers and resonant sensing elements, which are fabricated from silicon wafer by micro-machining process. The aim of the mechanical structure design for the seismic sensors is to achieve ultrahigh frequency sensitivity to acceleration input. The frequency-acceleration sensitivity for the MEMS RXL topology in Figure 6.2 is given by:

$$\frac{\Delta f_{out}}{a_{in}} = S_{Res} \times EA_{Lvr} \times M_{Proof} \tag{6.1}$$

where S_{Res} is the scale factor of the resonant sensing element in the unit of 'Hz/N', EA_{Lvr} the effective amplification factor of the micro levers, M_{proof} the proof-mass, a_{in} the input acceleration and Δf_{out} is the frequency shift of the RXL. In order to achieve high sensitivity, the parameters on the right-hand side of Equation (6.1) should be maximized. Meanwhile, the impacts of design parameters on other performance metrics, such as the dynamic range, bandwidth, mechanical robustness and constraints imposed by fabrication limitations, also need to be considered.

6.2.3 Resonant Sensing Element

The resonant sensing element serves two functions: one is to convert the inertial force change to a resonant frequency shift; the other is to transduce the sensed response into an electrical form. The second function is mainly related to the transduction scheme and electrical circuit design which will be discussed later in Section 6.3. However, the first function is directly related to the scale factor of resonant sensing element, S_{Res}, in Equation (6.1), which will be discussed in the next sub-section.

6.2.3.1 Analytical model

The resonant frequency of the sensing element is designed to be sensitive to an axial force load applied on it. Therefore, any resonator topology with such a characteristic can be used as the sensing element, such as cantilever resonator, single clamp-clamp beam resonator and FBAR resonator. For the RXL in this chapter, a Double-Ended Tuning Folks (DETF) topology is chosen for the sensing element design since it has high sensitivity to the axial load, high quality factor and simple topology. The DETF topology has been widely used in other resonant MEMS sensors as well.

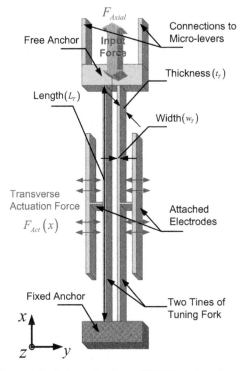

Figure 6.2 Schematic view of DETF sensing element.

Figure 6.2 provides a schematic view of the DEFT sensing element for the RXL. One anchor of the DETF is fixed, whereas the other is free connected to the micro-levers as the input port for the inertial force from the proof mass. The attached electrodes (see Figure 6.2) provide electro-mechanical transduction between the sensing element and the electrical circuits. For the purpose of mechanical analysis, they are only considered as attached masses on the tine centre. Assuming that there is no deflection at the two anchors (although later discussion will show this assumption is only applicable under certain condition), two fork tines can be viewed as separate clamped-clamped beams (C-C beam) with actuation forces along the length. The differential equation that describes the beam motion caused by the bending and tension effect is given by [11]:

$$\frac{\partial^2}{\partial x^2}\left(E\cdot\frac{t_T\cdot w_T^3}{12}\cdot\frac{\partial^2 d_T}{\partial x^2}\right)+\frac{\partial}{\partial x}\left(F_{Axial}\frac{\partial d_T}{\partial x}\right)+\rho\cdot t_T\cdot w_T\frac{\partial^2 d_T}{\partial t^2}=F_{Act}(x)$$

$$(6.2)$$

where $d_T(x, t)$ is the deflection of the beam, E is the modulus of elasticity of the material, ρ is the density of the material, L_T, w_T and t_T are the geometry dimension variables illustrated in Figure 6.2, and F_{Axial} and F_{Act} are the axial tension and the transverse force applied to the beam, respectively. The boundary conditions assume that the ends of the beam has no deflection or slope:

$$d_T(0) = d_T(L_T) = 0; \quad \frac{dd_T}{dx}\bigg|_{x=0} = \frac{dd_T}{dx}\bigg|_{x=L_T} = 0 \tag{6.3}$$

When the natural frequency of the beam is calculated, it is assumed that no transverse force is applied to the beam. So, $F_{Act}(x)$ is equal to zero at every position of the beam except where the attached masses are placed. The inertial force of each attached mass has to be taken into account:

$$F_{Act}(x) = \sum_j m_j \ddot{d}_T(x_j)\delta(x_j) \approx m_{Ele}\ddot{d}_T(L_T/2)\,\delta(L_T/2) \tag{6.4}$$

where m_{Ele} is the mass of an attached electrode, assuming that the position of this electrode is at the centre of beam.

In the analysis of the partial differential Equation (6.2), it is reasonable to assume that the responses of the beam, $d_T(x, t)$, include an infinite number of vibration modes which are orthogonally among each other, and each vibration mode response consists of a time-dependent and a position-dependent term, respectively:

$$d_T(x, t) = \sum_i q_i(t) \cdot \phi_i(x) \tag{6.5}$$

where ϕ_i is the i-th mode shape, and q_i is the associated modal coordinate. Substitute Equation (6.5) into Equation (6.2) and utilize the orthogonally relations between every two modes, the Equation (6.2) is simplified for the i-th mode and becomes:

$$\left(\int_0^{L_T} \rho \cdot t_T \cdot w_T \phi_i^2 dx + \int_0^{L_T} m_{Ele} \cdot \phi_i(L_T/2) \cdot \delta(L_T/2) \cdot \phi_i dx\right) \ddot{q}_i$$
$$+ \left(\int_0^{L_T} \left(E \cdot \frac{t_T \cdot w_T^3}{12}\frac{\partial^4 \phi_i}{\partial x^4} + F_{Axial}\frac{\partial^2 \phi_i}{\partial x^2}\right) \cdot \phi_i dx\right) q_i = 0 \tag{6.6}$$

The first term in the above equation can be seen as the inertia term, M_{eff}, and the second as the stiffness term, K_{eff}. The solution for the i-th-mode in the above second-order system is a harmonic motion with modal frequency given by [12]:

$$\omega_i = \sqrt{\frac{K_{eff}}{M_{eff}}}, \text{ where}$$

$$M_{eff} = \int_0^{L_T} \rho \cdot t_T \cdot w_T \cdot \phi_i^2 dx + m_{Ele} \cdot \phi_i^2 (L_T/2)$$

$$K_{eff} = \int_0^{L_T} E \cdot \frac{t_T \cdot w_T^3}{12} \left(\frac{\partial^2 \phi_i}{\partial x^2}\right)^2 dx + \int_0^{L_T} F_{Axial} \left(\frac{\partial^2 \phi_i}{\partial x^2}\right)^2 dx \quad (6.7)$$

According to Equation (6.7), the resonant frequency of the DETF sensing element is the function of input axial force, F_{Axial}. So, the scale factor of the resonant sensing element, S_{Res}, can be derived as follows:

$$S_{Res} = \frac{1}{2\pi} \cdot \frac{\partial \omega_i}{\partial F_{Axial}}$$

$$= \frac{1}{4\pi \omega_i M_{eff}} \frac{\partial K_{eff}}{\partial F_{Axial}} = \frac{1}{4\pi \omega_i M_{eff}} \times \int_0^{L_T} \left(\frac{d\phi_i}{dx}\right)^2 dx \quad (6.8)$$

Before the scale factor can be found from Equation (6.8), the only undetermined parameters are the vibration mode shapes ϕ_i. However, due to the presence of the axial force term, F_{Axial}, and the presence of the attached mass term, m_{Ele}, a closed-form solution for the mode shapes is not trivial [13]. Alternative approaches using energy methods can potentially achieve quite good approximations [11].

In the above analysis, the individual tines were treated as single beams. Although it can predict the modal frequencies and scale factor, the two tines of DEFT sensing element will interact between each other during the vibration. In each vibration mode, for the individual tines, there will be a pair of modes present in the DETF sensing element – two tines may vibrate in the symmetric mode or the anti-symmetric mode. Although, in principle, any mode of the sensing element can be used for sensing, the fundamental modes are preferred for several practical considerations, such as the ease designs of excitation/sensing transducers and interface electronics. Two fundamental mode shapes of the sensing element simulated by COMSOL® 4.2a are shown in Figure 6.3.

Since the stresses imposed by the two tines will cancel out when the two tines vibrate anti-symmetrically, the two anchors of the DETFs are not moving in the anti-symmetric mode, which agrees with the boundary conditions assumed in Equation (6.3). Therefore, the frequency of anti-symmetric mode

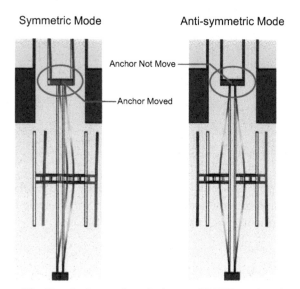

Figure 6.3   Two fundamental mode shapes of DETF sensing element.

can be estimated from Equation (6.7). However, the stress imposed by the two tines will not cancel out when they are vibrating symmetrically. The non-zero stress acts on the free anchor breaks the boundary conditions assumed in Equation (6.3). The symmetric mode shown in Figure 6.3 involves vibration coupling at the free anchor and the simple 'clamped-clamped, beam model descripted in Equations (6.2)–(6.7) must be combined with modelling of the vibration coupling at the anchor.

Compared to the symmetric mode of the sensing element, the anti-symmetric mode provides a higher quality factor than the other mode that benefits from lower anchor loss [12, 14]. Therefore, the mode of the DETF sensing element discussed in the remaining part of the chapter is the anti-symmetric mode unless being specified.

6.2.3.2 Scale factor optimization

This sub-section presents the mechanical geometry design optimization for the scale factor of the DETF sensing element. Since the fundamental anti-symmetric mode is selected for the operation of the sensing element, Equation (6.8) can be simplified by substituting the fundamental mode shape function of C-C beam and geometry terms, which is given as [15]:

$$S_{Res} = \frac{\Delta f}{F_{Axial}} \approx \frac{1}{4} S \cdot f_c, \text{ where } S = 0.293 \left(\frac{L_T^2}{E t_T w_T^3} \right)$$

$$\text{and } f_c = \frac{2}{\pi} \cdot \sqrt{E \left(\frac{w_T}{L_T} \right)^3 \cdot \frac{1}{\rho(A_{Ele} + 0.375 \cdot L_T w_T)}} \tag{6.9}$$

where A_{Ele} is the area of attached electrode shown in Figure 6.2, other variables are the same as Equation (6.8). As shown in Equation (6.9), the scale factor is determined by the material property and the geometry dimensions. Since the choice of material parameters are often constrained by the fabrication process (and taken to be single-crystal silicon in this work), the optimization studies will focus on the dimensions of DETFs sensing element. The scale factor is dependent related to the dimensions is derived from Equation (6.9) as below:

$$S_{Res} \propto \frac{L_T^{1/2}}{t_T w_T^{3/2} \left(A_{Ele} + C_2 \cdot L_T w_T \right)^{1/2}} \tag{6.10}$$

where C_2 is a constant determined by the material and the mode shape of DETF sensing element.

According to Equation (6.10), the scale factor of the sensing element can be increased by decreasing the width and thickness of the tuning fork tine. And the size of attached electrode also needs to be designed as small as possible to achieve high scale factor. Since the size of the attached electrode, A_{Ele} is not zero, increasing the tine length can also increase the scale factor to some extent. However, since the tine length term, L_T, is in both the numerator and denominator of Equation (6.10) and has the same order, increasing the tine length will not significantly improve the scale factor, comparing to decreasing the width, w_T, and thickness, t_T.

Using COMSOL® 4.2a, the influence of tine length, width, thickness and the attached electrode size to the scale factor can be simulated and studied. The reference geometry dimensions of DETFs sensing element in the simulation are as follows:

$$L_T = 350 \, \mu m$$
$$w_T = 4 \, \mu m$$
$$t_T = 30 \, \mu m$$
$$A_{Ele} = 1250 \, \mu m^2$$

The material model used in the simulation is single-crystal silicon. The scale factor estimation is derived from the pre-stressed eigenfrequency simulation results. The axial force load applied in the simulations is 1 μN- a small force to avoid mechanical nonlinearity of the sensing element. Figures 6.4 to 6.7 show the relationship of the scale factor to different tine widths, lengths, thicknesses and electrode sizes, respectively.

The simulation results shown in the above four graphs demonstrate the effects of each geometric dimension on the scale factor of the DETF sensing element, which agrees with the analytic modelling provided in Equation (6.9). The offset between the analytical model results and the FEM simulation results shown in Figures 6.5 and 6.7 are believed to result from the simplified mode-shape used in the analytical model. It is clear that the sensing element having thin, narrow and long tines and small electrodes will have a large scale factor. However, a sensing element may be limited by nonlinearity, shock robustness and fabrication yield issues, which will be discussed later.

6.2.3.3 Nonlinearity of scale factor

The modelling and analysis in this section so far assume that the scale factor is constant, regardless the direction and magnitude of input force. However, if the input force becomes large, the frequency shift of the DETF sensing element will exhibit deviation from linearity. For a large tensional input force,

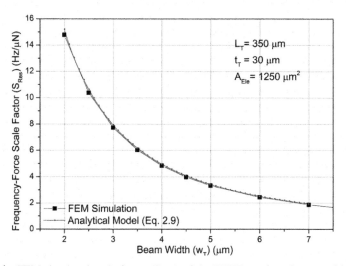

Figure 6.4   FEM simulated scale factor (S_{res}) of the DETF sensing element with different beam width.

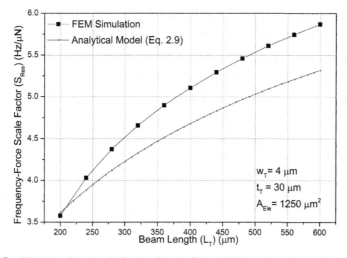

Figure 6.5   FEM simulated scale factor (S_{res}) of the DETF sensing element with different beam lengths.

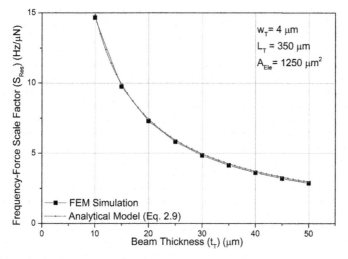

Figure 6.6   FEM simulated scale factor (S_{res}) of the DETF sensing element with different beam thicknesses.

the frequency shift will increase, whereas for a large compressive input force, the frequency shift will decrease. For the same input force, the DETF sensing element with thin, narrow tines will exhibit more nonlinearity in its frequency response than thick, wider tines. This nonlinear relation between the input

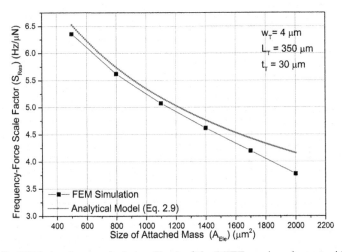

Figure 6.7 FEM simulated scale factor (S_{res}) of the DETF sensing element with different size of attached electrodes.

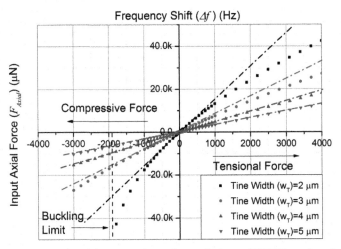

Figure 6.8 FEM simulated frequency-input force relations of DETF sensing element with different beam widths.

force and the frequency shift of the DETF sensing element is studied by numerical simulation and the results are shown in Figure 6.8. When the tine width decreases from 5 μm to 2 μm, the frequency shift notably deviated from the linear-relations to the input force. The asymmetry of the frequency shift between the compressive and tensional input force is also evident in the sensing element with narrower tine. The simulation results even indicate that the 2-μm tine may buckle under about 2 mN compressive input force.

6.2.3.4 Conclusions

This section introduces the operation principle of the DETF sensing element and derives the simplified analytic model for scale factor. Using numerical simulations, the analytic model is verified and the geometry influences the scale factor nonlinearity of the sensing element are studied. Apart from the scale factor, the mechanical design must also address operation frequency, power handling ability, resolution limitation, maximum shock load, and other requirements, which will be discussed in other sections.

6.2.4 Inertial Force Amplifier

The micro-lever is a critical component to enable the increased sensitivity for the MEMS RXL. It amplifies the inertial force of proof-mass and couples the amplified force to the DETF sensing element. This section presents the design and optimization of the structure and geometry of the micro-lever to maximize its effective amplification factor, EA_{Lvr}, in Equation (6.1).

6.2.4.1 Micro-leverage mechanism

The basic principle of the micro-lever force amplifier is based on the leverage mechanisms as described by the Greek scientist and philosopher Archimedes. However, the structure of the micro-lever is different from the conventional macro levers. Because of the constraints of the micro-machining fabrication process, it is difficult to implement the 'hinge' as the pivot in the micro-lever. Thus, the micro-leverage mechanism is mainly formed by coplanar flexures, with one end of the flexure beam anchored to the substrate as a pseudo-pivot. Mechanical transformation in a micro-leverage mechanism is achieved by elastic deformation of its component flexure beams.

Similar to the conventional leverage mechanism, the micro-lever may amplify the input force or displacement and provide inverted or non-inverted output. Figure 6.9 shows four kinds of single-stage micro-levers with different functions, respectively. These micro levers have been used as either force amplifier or displacement amplifier in a number of MEMS devices, including the capacitive accelerometer [16] as well as the resonant accelerometer [17].

According to the application specified by the RXL and the constraint of micro-machining fabrication, the micro-lever force amplifier with inverse output is used in the MEMS RXL. However, the design and analysis methods

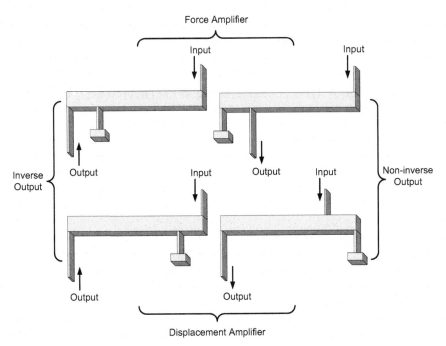

Figure 6.9   Four Kinds of Micro-Lever with Different Functions [18].

presented in the following sections are also applicable to other kinds of micro-levers shown in Figure 6.9.

6.2.4.2 Lever amplification factor

The single-stage micro-lever consists of an input beam, a lever beam, a pivot beam, a pivot anchor and a connection beam. The input beam couples the inertial force of the proof-mass onto the micro-lever. The lever beam, pivot beam and the pivot anchor mimic the conventional lever structure. The connection beam passes the amplified force to the DETF sensing element. Figure 6.10 shows the schematic view of a single-stage micro-lever force amplifier. Lever amplification factor, A_{Lvr}, is defined as the ratio of the output force magnitude to the input force magnitude of a micro-lever, which is different from the effective amplification factor, EA_{Lvr}, in Equation (6.1). Lever amplification factor is mainly determined by the structure and geometry design of a micro-lever.

Figure 6.11 shows the first-order mechanical model of single-stage micro-lever. The model assumes that the lever beam is rigid and it also ignores the

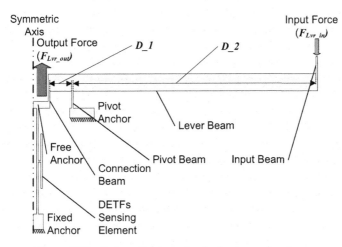

Figure 6.10 Schematic view of a single-stage micro-lever.

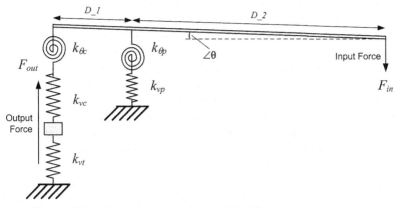

Figure 6.11 First-order mechanical model of single-stage micro-lever.

connection area between lever beam and other beams, and the horizontal force caused by the bending of the input beam. As opposed to a conventional hinge, the flexural beam has non-zero rotational stiffness and finite vertical axial stiffness, which are represented by k_θ and k_ν in Figure 6.11, respectively. The subscripts 'p', 'c' and 't' represent pivot beam, connection beam and tine beam, respectively. The rotational stiffness of the tine beam is omitted in the model since the two tuning fork tine beams are bended towards the opposite directions by a pair of micro-levers ands the bending force on the beams will be cancelled by the symmetric structure of the DETF sensing element.

The vertical axial stiffness and rotational stiffness of single ended flexure beam are given in Equations (6.11) and (6.12):

$$k_v = \frac{E w_B t_B}{L_B} \tag{6.11}$$

$$k_\theta = \frac{4 E I_B}{L_B} = \frac{E t_B w_B^3}{3 L_B} \tag{6.12}$$

where the w_B, t_B and L_B are the width, thickness and length of the flexure beam, respectively. The lever amplification factor can be estimated from the mechanical model (as shown in Figure 6.11) as follows:

According to the definition, the lever amplification factor is:

$$A_{Lvr} = \frac{F_{Lvr\_out}}{F_{Lvr\_in}} \tag{6.13}$$

Assuming the moment and force of the micro-lever are in balance and solve the equation for F_{Lvr_in} and F_{Lvr_out}:

$$\begin{cases} F_{Lvr\_in} = \frac{\theta}{\cos\theta} \left(\frac{k_{\theta c}}{D\_1 + D\_2} + \frac{k_{\theta p}}{D\_2} \right) + \frac{D\_1}{D\_2} \cdot F_{Lvr\_out} \\ F_{Lvr\_in} = F_{Lvr\_out} \cdot \left(1 + \frac{k_{vp}}{k_{vc}} + \frac{k_{vp}}{k_{vt}} \right) - \sin\theta \cdot k_{vp} D\_1 \end{cases} \tag{6.14}$$

Substituting the solutions of Equation (6.14) into Equation (6.13):

$$A_{Lvr} = \frac{C_1 + \sin\theta \cdot k_{vp} D\_1}{C_1 \left(1 + \frac{k_{vp}}{k_{vc}} + \frac{k_{vp}}{k_{vt}} \right) + \sin\theta \cdot k_{vp} \frac{D\_1^2}{D\_2}},$$

$$\text{where } C_1 = \frac{\theta}{\cos\theta} \left(\frac{k_{\theta c}}{D\_1 + D\_2} + \frac{k_{\theta p}}{D\_2} \right) > 0 \tag{6.15}$$

According to Equation (6.15), all the parameters shown in Figure 6.11 will influence the lever amplification factor. Hence, there are two design strategies to increase the lever amplification factor: 1) increase the lever ratio and 2) decrease the rotational stiffness. In addition to Equation (6.15) and the simplified mechanical model shown in Figure 6.11, it will be helpful to use FEM simulation (using COMSOL® 4.2a) to evaluate the above two optimization methods as well, since it can take into consideration many simplified factors in the first-order mechanical model, for instance, the influence of the flexural rigidity of the lever beam. The material model used in the simulation is single crystal silicon. The default geometry dimensions of the micro-lever model

Table 6.1 Initial Geometry Dimensions of Micro-Lever Model for Amplification Factor Simulation

Name of Component	Dimensions (μm)
Lever Beam	2480 (L) $\times$ 170 (w) $\times$ 30 (t)
Pivot Beam	50 (L) $\times$ 5.5 (w) $\times$ 30 (t)
Connection Beam	45 (L) $\times$ 4 (w) $\times$ 30 (t)
Input Beam	200 (L) $\times$ 3 (w)$\times$ 30 (t)
Tine of DETFs	280 (L) $\times$ 3 (w) $\times$ 30 (t)
D_1	38.5
D_2	$2480-(3+4+5.5)-D\_1$

used in simulation are listed in Table 6.1, where the component names are same as those in Figure 6.10.

- **Increasing the lever ratio**

Similar to the conventional lever, the lever amplification factor will increase, while the pivot beam is moved closer to the output beam. However, since the rotational stiffness of connection beam and pivot beam do not equal to zero, it can be shown that the lever amplification factor is always smaller than the ideal lever ratio, $A_{Ideal} = \frac{D\_2}{D\_1}$ as below:

$$A_{Lvr} = \frac{C_1 + \sin\theta \cdot k_{vp}D\_1}{C_1\left(1 + \frac{k_{vp}}{k_{vc}} + \frac{k_{vp}}{k_{vt}}\right) + \sin\theta \cdot k_{vp}\frac{D\_1^2}{D\_2}}$$

$$< \frac{C_1 + \sin\theta \cdot k_{vp}D\_1}{C_1 + \sin\theta \cdot k_{vp}\frac{D\_1^2}{D\_2}} \qquad (6.16)$$

$$\because D\_1 < D\_2 \Rightarrow D\_1 > \frac{D\_1^2}{D\_2}$$

$$\therefore \frac{C_1 + \sin\theta \cdot k_{vp}D\_1}{C_1 + \sin\theta \cdot k_{vp}\frac{D\_1^2}{D\_2}} < \frac{\sin\theta \cdot k_{vp}D\_1}{\sin\theta \cdot k_{vp}\frac{D\_1^2}{D\_2}} = \frac{D\_2}{D\_1}$$

$$\therefore A_{Lvr} < A_{Ideal} \qquad (6.17)$$

Another limitation of this method is that the rotary angle of lever beam, $\angle\theta$, will increase with the increased ideal lever ratio under same input force. This C_1 in Equation (6.15) will also increase and it may partially or even fully cancel the benefit of the increased lever ratio to the lever amplification factor.

Figure 6.12 shows the comparison of the simulated lever amplification factor with the ideal lever ratio when the position of the pivot beam changes. The simulation results show that the lever amplification factor, A_{Lvr}, is always smaller than the ideal lever ratio because of the energy loss inside

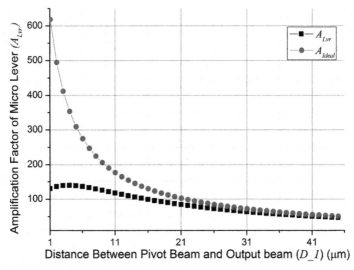

Figure 6.12   Comparison of FEM simulated lever amplification factor to ideal lever ratio.

the micro-lever, which agrees with the analysis presented earlier. Part of the energy loss could result from the non-zero rotational stiffness and finite vertical axial stiffness of the beams, which is modelled by Equation (6.15). The bending of the lever and input beams may also result in an energy loss. While the pivot beam is moved close to the connection beam, the lever amplification factor deviated further and further away from the ideal lever ratio and the lever amplification factor starts decreasing when the distance between pivot beam and connection beam is less than 4 μm.

• Decreasing the rotational stiffness

As shown in Equation (6.16), if the rotational stiffness equals zero, the lever amplification factor will increase and approach the ideal lever ratio. Even though the rotational stiffness of connection beam and pivot beam are always non-zero, reducing them could also increase the lever amplification factor. According to Equation (6.12), a long, thin and narrow beam has a relatively low rotational stiffness. However, the beam with such a geometry also has low vertical axial stiffness (see Equation (6.11)), which will decrease the lever amplification factor. Comparing Equation (6.11) and Equation (6.12), it can be seen that only reducing the beam width can make the rotational stiffness decrease more quickly (cubic dependency of width) than the vertical axial stiffness (linear dependency of width).

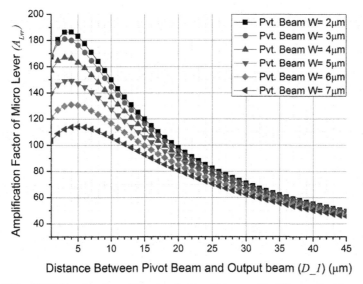

Figure 6.13 FEM Simulated amplification factor of levers with different pivot beam width.

Figures 6.13 and 6.14 show the comparison of a simulated lever amplification factor of the micro-levers with different widths of pivot beam and connection beam, respectively. Micro-levers with a narrow pivot beam or connection beam have a larger amplification factor than the micro-levers with a wide beam, which agrees with the above analysis. The simulation results also indicate that the influence of rotational stiffness to lever amplification factor becomes more noticeable with the increase of lever ratio. Therefore, for those micro-levers with large lever ratios, reducing the width of the pivot beam and connection beam could remarkably enlarge the lever amplification factor. However, the width of the beam cannot be designed infinite small. The minimum beam width in a design is normally limited by the fabrication process.

All the above discussions of the micro-lever amplification factor are based on the single-stage lever topology. However, micro-levers with two or more stages may be useful in some cases. Especially when the device area is not sufficient to accommodate a long lever beam to achieve a large lever ratio using single-stage topology, the multi-stage lever topology can be adopted [17]. However, since the multi-stage micro-lever consists of more than one pivot beam and connection beam, the energy loss in the multi-stage

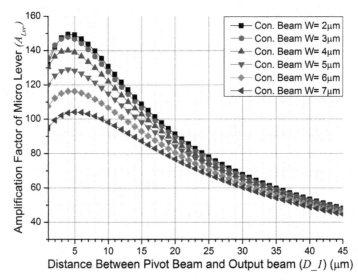

Figure 6.14   FEM Simulated amplification factor of levers with different connection beam widths.

micro-lever might be larger than the energy loss in the single-stage micro-lever. That means the lever amplification factor of a multi-stage micro-lever could be smaller than a single-stage micro-lever while they have same lever ratio.

6.2.4.3 Effective amplification factor

The effective amplification factor of a micro lever is defined by Equation (6.18), which is used to evaluate the micro-lever performance in the MEMS RXL.

$$EA_{Lvr} = \frac{F_{Lvr\_out}}{M_{Proof}a_{in}} \qquad (6.18)$$

By comparing Equation (6.18) to Equation (6.13), it can be noticed that if the inertial force of the proof mass fully couples to the micro-lever, the effective amplification factor (EA_{Lvr}) will equal to the lever amplification factor (A_{Lvr}). However, as shown in Figure 6.1, the proof mass of MEMS RXL is supported by suspensions that have non-zero flexure stiffness on the sensing axis. Because of the flexure stiffness of suspensions (k_{sus}), the deformed suspensions will partially balance the inertial force of the proof mass, while the proof-mass is moved under the acceleration input, which will reduce the inertial force coupled to the micro-lever and make the effective amplification factor smaller than the lever amplification factor. The effective amplification factor of a micro-lever with suspensions can be estimated as below,

$$EA_{Lvr} = \frac{F_{Lvr\_out}}{M_{Proof}a_{in}} = \frac{A_{Lvr}F_{Lvr\_in}}{M_{Proof}a_{in}} \qquad (6.19)$$

where the A_{Lvr} is lever amplification factor of the micro-lever. By defining the input equivalent stiffness of the micro-lever as,

$$k_{Lvr\_in} = \frac{F_{Lvr\_in}}{x_{Lvr\_in}} \qquad (6.20)$$

where x_{Lvr_in} is the displacement of the input end of the micro-lever, and assuming that the displacements along sensing axis of the proof mass, free end of the suspensions and the input beam of the micro-lever are all the same, Equation (6.19) can be rewritten as:

$$EA_{Lvr} = \frac{k_{Lvr\_in} \times A_{Lvr}}{k_{sus} + k_{Lvr\_in}}. \qquad (6.21)$$

By ignoring the beam deformation, the input equivalent stiffness of the micro lever can be approximated by the lever amplification factor and vertical axial stiffness of the tine of DETFs sensing element:

$$k_{Lvr\_in} \approx \frac{k_{vt}}{A_{Lvr}^2} \qquad (6.22)$$

Substitute Equation (6.22) into Equation (6.21), the effective amplification factor can be expressed as a function of the lever amplification factor:

$$EA_{Lvr}(A_{Lvr}) = \frac{k_{vt}A_{Lvr}}{k_{sus}A_{Lvr}^2 + k_{vt}} \qquad (6.23)$$

Then the maximum effective amplification factor is:

$$Max(EA_{Lvr}) = \frac{A_{Lvr}}{2}, \text{ when } A_{Lvr} = \sqrt{\frac{k_{vt}}{k_{sus}}} \qquad (6.24)$$

According to Equation (6.24), the effective amplification factor is always smaller than or equal to half of the lever amplification factor when the suspensions are considered. Equations (6.23) and (6.24) also indicate that for a given design of DETFs sensing element and suspensions where k_{sus} and k_{vt} are fixed, there will be an optimum lever amplification to maximize

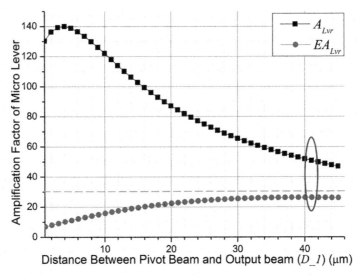

Figure 6.15   FEM simulation results of the lever amplification factor and the effective amplification factor.

the effective amplification factor. Figure 6.15 shows the simulated lever amplification factor (A_{Lvr}) and effective amplification factor (EA_{Lvr}) of the same micro-lever.

The results from the analytical model and numerical simulation described above indicate that the effective amplification factor of the micro-lever in MEMS RXLs is not only determined by the lever amplification factor, but also related to the suspensions and DETF sensing element design. The effective amplification factor of the micro lever shown in Figure 6.15 has a maximum value of 26, which equals to half of the lever amplification factor of 52. and the effective amplification factor starts decreasing with larger lever ratio even though the lever amplification factor still increases, which agrees with the analysis. However, according to Equation (6.24), the effective amplification factor of micro-lever can be further increased by lowering the stiffness of suspensions (k_{sus}) along the sensing axis.

6.2.4.4 Conclusion

The purpose of using micro-levers in MEMS RXLs is to achieve high acceleration sensitivity with a relatively small design area. Although the lever amplification factor represents the force amplification ability of the micro-lever, it is shown that the force amplification performance of a micro-lever is not only dependent on the lever amplification factor, but also related to the design of suspensions and DETF sensing element. Therefore, the design of the micro-lever must obtain an optimum lever amplification factor within the smallest design area for a given designs of suspensions and DETFs sensing element, in order to improve the performance of RXL.

Compared to the single-stage micro-levers, a multi-stage micro-lever could potentially achieve a higher lever amplification factor, but its energy loss is also larger than the single-stage micro-lever with same lever amplification factor. This means that the single-stage topology is preferred in the micro-lever design unless the optimum lever amplification factor cannot be achieved by a single-stage micro-lever.

6.2.5 Proof Mass and Suspension Frame

In an MEMS RXL, the input acceleration is converted to the inertial force of the proof-mass, which is further amplified by a micro-lever and measured by a DETF sensing element. Since the RXL needs to distinguish three axial components of the input acceleration, the suspensions connected to proof-mass are designed to pass through the inertial force component along the sensing axis but reject all other inertial force components of the proof-mass.

6.2.5.1 Proof-mass design

The main design parameter of the proof-mass is the weight or the volume that directly relates to the sensitivity of RXL as represented in Equation (6.1). A large proof-mass is preferred in the design for high sensitive MEMS RXLs. Because of the characteristic of the micro-machining fabrication process, the thickness of the device is normally fixed and relatively thin. The large proof-mass needs to be designed like a membrane structure where the length and width are much larger than the thickness. However, such a membrane-like proof-mass could be deformed by self-weight. Figure 6.16 shows the numerical simulation result of a 5 mm × 5 mm × 30 μm (which is corresponding to specific fabrication process) proof-mass whose suspensions along the side are bended down by its weight. This unwanted deformation of proof-mass may cause stiction issues if it is brought in contact with the substrate and it could also induce more stress to the micro-lever and suspensions, which might reduce the robustness of RXL.

There are two methods that can reduce the self-deformation of the proof-mass. One is to increase the thickness to make the proof-mass more rigid. The other is to optimize the position of the suspensions. Since the device thickness is fixed or can only be changed in a certain range in most micro machining fabrication processes, the application of the first method from

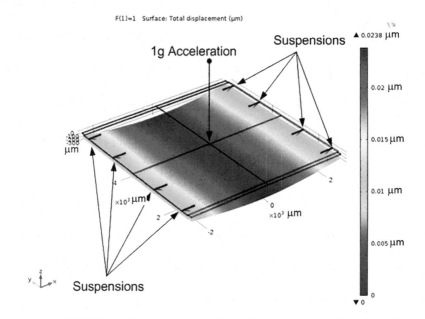

Figure 6.16   FEM Simulation of large membrane-like proof-mass self-deformed by its weight.

a design perspective is limited. The second method does not change the stiffness of proof-mass, but could allow uniform displacement of the proof-mass under acceleration load and thus reduces the proof-mass deformation. As shown in Figure 6.17, four suspensions are moved from two sides to the centre of proof-mass. Comparing Figure 6.17 to Figure 6.16, it can be noticed that the maximum displacement of proof-mass reduced by more than 80% by just changing the positions of suspensions. However, the application of the second method also has limitations since not every fabrication process allows an anchor to be placed in the centre of a suspended structure.

Although a large proof-mass could increase the sensitivity of the MEMS RXL, the size of the proof-mass is not always the larger the better. Since the proof-mass also relates to the mechanical bandwidth, dynamic range, shock resistance and fabrication yield of MEMS RXL, the design of the proof-mass will be a trade-off among all the above factors. For example, a larger proof-mass weight will result in a smaller mechanical bandwidth and a dynamic range of the MEMS RXL as well as attenuate the shock resistance and make the fabrication process more challenging.

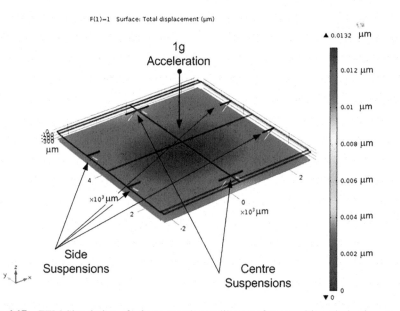

Figure 6.17 FEM Simulation of a large membrane-like proof-mass with optimized suspensions and deformed by its own weight.

6.2.5.2 Suspensions design

The main design parameters of the suspensions are the flexure stiffness of the three axes. An ideal suspension system should have zero stiffness in the sensing axis and an infinite stiffness in other axes. In practice, the suspensions should be flexible along the sensing axis but rigid along the other axes. As discussed in the section of the micro-lever design, the suspension stiffness in the sensing axis influences the performance of the micro-lever and consequentially the sensitivity of RXL.

In principle, it is not necessary for the MEMS RXL to have an independent suspension structure, as the micro-lever could also partly possess the function of suspensions. However, a special designed suspension structure is very important in the design of a high-sensitivity RXL, because the large proof-mass size and the long lever beam make the out-of-plane stiffness of the micro-lever too low to be used as the suspensions of the proof-mass.

Since many MEMS devices with a movable structure use suspensions, there are a number of suspension topologies that could potentially be used for an MEMS RXL. However, the nonlinearity of the suspension stiffness is an important factor that needs to be considered here. For example, the flexure stiffness of single beam can be approximated to a constant only for the small deformation of the beam. Axial tensile stress is present in laterally deflected single beam suspension. A nonlinearity of the force-displacement of the single beam results from this axial tensile stress, where the effective stiffness constant increases with the increase of the force load. As described by Equation (6.24), an RXL with highly sensitive DETF sensing element design would prefer low transverse stiffness of the suspension to increase the effective lever amplification factor. Since the increase of lever amplification factor also increases the displacement of suspensions, the nonlinearity associated with large deflection of suspension may offset the increase of lever amplification factor.

6.2.6 Mechanical Structure Design Evaluation

After the study of the design methodology of each component in the mechanical structure of RXL, the methods for assembling and evaluation of a complete mechanical structure design will be introduced in this section.

The mechanical structure assembly of an RXL is determined by the design targets, which may include but not limited to the sensitivity along the desired input axis, scale factor, bandwidth, input acceleration range, robustness and size. The number of sensitive axes determines the choice of suspension topology, where the single beam or snake beam suspension can be used for single-axis RXL and parallel kinematic beams suspension is for dual-axis RXL. The scale factor is related to all of the components (see Equation (6.1)) studied in this section before. To achieve high scale factor, a large proof-mass, sensitive DETFs with narrow tines, a micro-lever with a large lever ratio and low stiffness suspension will be assembled for an RXL. However, the scale factor-enhanced design may result in limited bandwidth, small acceleration input range and low robustness. Therefore, trade-offs between the scale factor and other characteristics must be made to meet the design targets. The size restriction of an RXL design normally limits the total area available for the mechanical structure assembly, which is another factor to be considered in the design trade-offs.

A design example of a single-axis RXL, as shown in Figure 6.18, will be studied in this section. The mechanical assembly process, FEM simulation methods and results are presented. In order to evaluate a design of RXL, FEM modelling and simulations by COMSOL® 4.2a are carried out for each mechanical assembly. The model of each design is built in 3-D with the same dimensions as the actual device. The anisotropic model of single

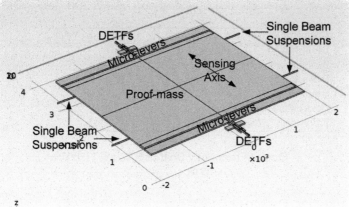

Figure 6.18   3D model of single-axis MEMS RXL.

Table 6.2   Summary of FEA simulation methods for the MEMS RXL characterization

Characteristic of RXL	Simulation Study Method
Scale Factor ($\frac{\Delta f_{out}}{a_{in}}$)	Pre-stressed Eigenfrequency study
Mechanical Bandwidth (BW_{Mech})	Stationary study with Harmonic Perturbation
Input acceleration range (a_{in_Max})	Pre-stressed Eigenfrequency study with Parameter Swipe
Robustness (a_{load_Max})	Linear Buckling and Stationary study

crystalline silicon is chosen for the simulation. Table 6.2 summaries the simulation methods used to evaluate the RXL design.

As shown in Figure 6.18, this single-axis RXL consists of two identical DETFs and micro-levers, which give the differential measurement of the input acceleration on the sensitive axis. The single-beam suspensions are used to minimize the out-of-plane deflection of the proof-mass. Table 6.3 summaries the dimensions of this RXL design.

Using the MEMS Module of COMSOL 4.3a, the performance of the RXL can be evaluated. As shown in Figure 6.19, the scale factor of the RXL is 143.5 Hz/m/s^2, and the linear range ($< 1\%$ linearity error) is more than $+/- 3$ g according to pre-stressed eigenfrequency simulation. Figure 6.20 shows the simulation results of the scale factor of the RXL versus the input acceleration frequency. Since the resonant frequency of the first in-plane mode of the prototype is about 950 Hz, the MEMS RXL is expected to

Table 6.3 Summary of geometry design of single-axis MEMS RXL with single beam suspensions

Geometry Dimensions	Designed Value
Beam width of DETF (w_{DETF})	3 μm
Beam length of DETF (L_{DETF}	350 μm
Beam width of Suspension (w_{Single})	3.5 μm
Beam length of Suspension (L_{Single})	400 μm
Device thickness (t)	25 μm
Proof-mass (M_{Proof})	468.36 μg

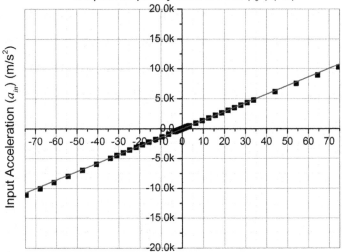

Figure 6.19 FEM Simulated scale factor and input acceleration range of the single-axis MEMS RXL.

have linear responses for input accelerations in the frequency range DC-100 Hz. Limited by the beam buckling and fracture stress of silicon [19], the maximum in-plane load cannot exceed 70 g (see Figure 6.21) and the maximum out of plane load cannot exceed 400 g (see Figure 6.22). But it should be noticed that in reality, the maximum in-plane/out of plane load of the MEMS RXL may be lower than the above values because of the fabrication defects as well as assembly or package issues. Table 6.4 summaries the main characteristics of the single-axis RXL with single-beam suspensions.

In summary, although the mechanical assembly of the RXL may vary with specific application requirements, the 3D modelling and FEM simulations

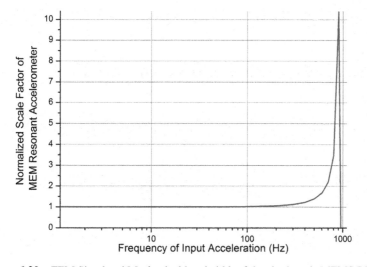

Figure 6.20   FEM Simulated Mechanical bandwidth of the single-axis MEMS RXL.

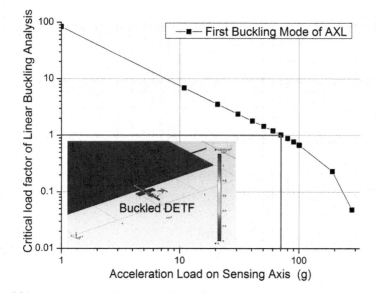

Figure 6.21   Linear buckling analysis of the mechanical design of single-axis MEMS RXL. (Inset: the positions where RXL buckled).

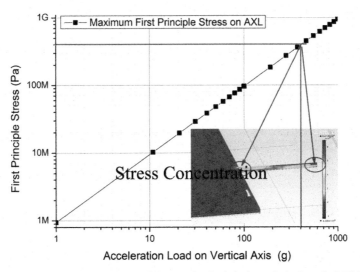

Figure 6.22   Fatigue stress analysis of the mechanical design of single-axis MEMS RXL (Inset: the positions where stress concentrated on the RXL).

Table 6.4   Summary of the characteristics of single-axis MEMS RXL obtained or derived from the simulations

Characteristics of RXL	Number Derived from Simulation
DETF resonant frequency (f_c)	111.49 kHz
Effective Amplification factor (EA_{Lvr})	25
Scale Factor ($\frac{\Delta f_{out}}{a_{in}}$)	143.5 Hz/m/s^2
Mechanical Bandwidth (BW_{Mech})	DC-100 Hz
Input acceleration range (a_{in_Max})	$+/-$ 30 m/s^2
Robustness (a_{load_Max})	70 g (In-plane)
	400 g (Out of plane)

can be used to evaluate the mechanical design and predict certain key characteristics of the RXL.

6.3 Electronic Circuitry Design

This section deals with the electronics design of MEMS RXL, which translates the mechanical vibration of the DETF sensing element to an electronic signal, tracks the resonant frequency change of DETF sensing element and demodulates the input acceleration from the frequency modulated signal. The electronic circuitry for the MEMS RXL consists of an electro-mechanical frequency tracking oscillator plus a frequency demodulator.

In this section, the design of the electro-mechanical oscillator will be described. The oscillator is composed of one DETF sensing element, electro-mechanical transducers, sustaining amplifier and other supplementary circuit components. The electro-mechanical transducer design and electro-mechanical model for the DETF sensing element are studied at first and then the electronic circuit design of oscillator will be introduced.

6.3.1 Electro-mechanical Transducer Design For DETF Sensing Element

The mechanical design of the DETF sensing element has been discussed in the previous sections. However, in order to make the DETF interact with the electronic circuits in the oscillator, an electro-mechanical transducer that converts the electrical power in the form of voltage and current to the mechanical power in the form of force and velocity, or vice versa is necessary. Electrostatic transducers are one of the most common examples of transducers used in MEMS devices. A capacitor is formed by one stationary and one movable part. When a DC voltage, referred to as 'polarization voltage', is applied across the capacitor, a charge is stored on the capacitor, and a force is generated between the two plates of the capacitor. If the voltage is increased, the mechanical force between the two plates increases, too. In contrast, if the parts move apart, the capacitance and therefore the stored charge decrease and result in a current flow. The voltage-to-force and motion-to-current relations can be derived by treating the structure as a mechanically variable capacitor. For a given voltage placed on the capacitor, the amount of energy stored in the system is:

$$U = \frac{1}{2}CV_0^2 \tag{6.25}$$

Where C is the capacitance. If assumed that the electrostatic transducer is lossless, the force exerted across the parts is:

$$P = -\frac{\partial U}{\partial x} = -\frac{1}{2}\frac{\partial C}{\partial x}V_0^2 \tag{6.26}$$

Normally, the resonator is driven by a voltage signal with a DC polarization voltage and an AC drive voltage at the same frequency as the resonator natural frequency, such as:

$$V = V_p + v_d \sin \omega t \tag{6.27}$$

Because the amplitude of AC signal is much smaller than the DC signal and the resonator only has significant response to the drive force with the frequency near its resonant frequency, the Equation (6.26) can be linearized to:

$$P = \frac{\partial C}{\partial x} V_p v_d \sin \omega t \tag{6.28}$$

The motion-to-current transduction factor can be derived from the change of the amount of charge stored across the capacitor:

$$Q = C V_0 \tag{6.29}$$

When one electrode of the capacitor moves, a current is generated caused by the change of stored charge across the two electrodes. The expression of the current can be obtained from the derivative of the above equation:

$$\frac{dQ}{dt} = i_s = \frac{dC}{dt} V_0 \tag{6.30}$$

where i_s is the sense current generated by the motion. Generally, the entire movable portion of the variable capacitor will move as one piece, thus the above equation can be broken down to:

$$i_s = \frac{dC}{dt} V_0 \frac{dx}{dt} \tag{6.31}$$

In the above equation, the sense current is proportional to the polarization voltage, V_0, the velocity of moving electrode, dx/dt, and the rate of capacitance change per unit electrode moving distance, dC/dx, which is determined by the transducer structure topology. The comb drive and parallel-plate are the most common electrostatic transducer topologies in MEMS. Equations (6.32) and (6.33) describe the rate of capacitance change with respect to the electrode displacement in a comb drive transducer and a parallel-plate transducer, respectively.

$$\frac{\partial C_{Comb}}{\partial x} = \frac{N \varepsilon \cdot t}{g_{elc}} \tag{6.32}$$

$$\frac{\partial C_{pal}}{\partial x} = \frac{\varepsilon \cdot t \cdot L_{etc}}{(g_{elc} - x)^2} \tag{6.33}$$

where N is the number of comb fingers, ε is the dielectric coefficient, g_{elc} is the initial distance between two comb fingers or two parallel-plate electrodes, L_{elc} is the length of electrode and t is the device thickness, which is also the height of the electrode. The advantage of the comb-drive transducer is that the capacitance variation of the comb electrode is independent of electrode position (see Equation (6.32)). However, the large electrodes of the

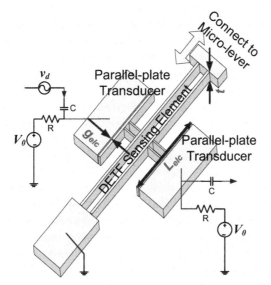

Figure 6.23 Schematic view of DETF sensing element with parallel-plate transducer.

comb-drive transducer attached on DETF will decrease the force-frequency scale factor of DETF sensing element (see Equation (6.10)). In contrast, the parallel-plate transducer requires much smaller electrode than the comb drive transducer to achieve the same electro-mechanical transduction efficiency, which is a significant advantage for DETF sensing element in the MEMS RXL. Although the relations between the capacitance and electrode position in the parallel-plate transducer is not linear (see 6.33), this nonlinearity may be attenuated by limiting the displacement of electrodes. So parallel-plate transducers are used for the sensing elementin this design, as shown in Figure 6.23.

6.3.2 Electro-mechanical Model of DETF Sensing Element

An electromechanical model is necessary to understand how the DETF sensing element interacts with the oscillation circuitry. Both linear and nonlinear models of the DETF sensing element will be discussed in this section.

6.3.2.1 Linear model of DETF sensing element

An electrical resonant tank is represented by a series or parallel LCR circuit. Thus, an equivalent electrical LCR circuit can be used to capture the behavior

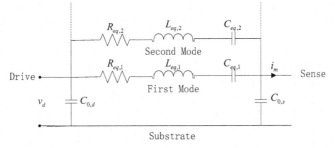

Figure 6.24   Equivalent electrical circuit for DETF actuated and sensed on both Tines [21].

of the DETF sensing element and model the linear vibration of the anti-symmetric mode. The conversion factors as well as the input and output currents from each electrode can be derived. Compared with the transfer function model [20], the equivalent electrical circuit is more straightforward for electrical simulation and circuit design. It not only can be used in simulation by circuit analysis tools such as SPICE, but also allows the analysis of the parasitic effects arising from other circuits and fabrication process. Figure 6.24 shows an equivalent electrical circuit for the DETF sensing element.

Similar to the LCR electrical circuit, the transfer function from the drive voltage of DETF resonator to the sense current is given below:

$$i_m = \sum \left(\frac{\frac{1}{L_{eq}}s}{s^2 + \frac{R_{eq}}{L_{eq}}s + \frac{1}{L_{eq}C_{eq}}} \right) v_d \tag{6.34}$$

where the equivalent electrical components are as follows:

$$L_{eq} = \frac{M_{eff}}{2K_PK_I} \tag{6.35}$$

$$C_{eq} = \frac{2K_PK_I}{M_{eff}\omega_c^2} \tag{6.36}$$

$$R_{eq} = \frac{M_{eff}\omega_c}{2QK_PK_I} \tag{6.37}$$

where M_{eff} is the effective mass of DETF with attached electrodes (see Equation (6.37)), ω_c is the angular frequency of the first out-phase mode of DETF (see Figure 6.3), Q is the quality factor of DETF resonator, K_p and

K_I are the electro-mechanical coupling coefficients on the driving and sensing ports, respectively. For an identical symmetric parallel-plate transducer design, these two coefficients are the same:

$$K_P = K_I = \frac{\varepsilon \cdot t \cdot L_{elc}}{g_{elc}^2} V_0 \qquad (6.38)$$

Large electro-mechanical coupling coefficients are normally preferred in the oscillator design to lower the motional resistance of DETF resonator and relax the design requirements for the sustaining amplifier. Since the electro-mechanical coupling coefficients are determined by the electrode dimensions and the DC polarization voltage (see Equation (6.38)). The coefficient can be increased by enlarging the electrode area, narrowing the gap between electrodes and increasing the DC polarization voltage. However, increasing the electrode area may decrease the scale factor of DETFs (see Equation (6.10)) and, on the other hand, high DC polarization voltage may result in the pull-in of two electrodes [12]. Because the above trade-offs, minimizing the gap between the electrodes (see Figure 6.23) is normally preferred in the parallel-plate transducer design. The minimum achievable gap is usually limited by the etching process in the fabrication. The $C_{0,d}, C_{0,s}$ are the static capacitances between the drive and sense electrodes and the resonator structure.

Since there are some parasitic effects from interconnect and packaging of the device, it is useful to add parasitic capacitance, $C_{p,d}$, $C_{p,s}$, C_{ft}, and series resistance R_s into the above model, as shown in Figure 6.25:

The parasitic resistance in the series with the resonator will reduce the quality factor, Q, of the DETF. The parasitic capacitance $C_{p,d}$ and $C_{p,s}$

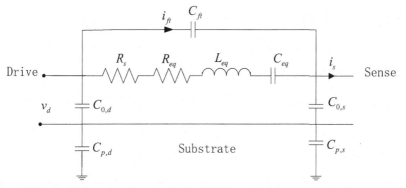

Figure 6.25 Equivalent electrical circuit for DETF sensing element actuated and sensed on both tines and with parasitic effect (1st Mode).

will couple additional noise into the DETF sensing element. And the feed-through capacitance, C_{ft}, may attenuate the output signal on the sensing port of DETF sensing element and result in an unwanted parallel resonant in the system. These parasitic capacitances can be reduced by increasing the distance between drive and sense port in the design, grounding the substrate and using specific drive/sense mode.

6.3.2.2 Nonlinear model of DETF sensing element

All the above analyses are based on the linear assumption of DETF sensing element. However, in practice, the nonlinearities of DETF sensing element may occur and will impact on the RXL performance. Therefore, a nonlinear model will be useful to understand the nonlinear response of DETF sensing element and evaluate its influence on the performance of RXL. The nonlinearity of the DETF sensing element can be divided into mechanical nonlinearity and electrostatic actuation nonlinearity, which will be described below.

• **Mechanical Nonlinearity**

The dominant mechanical nonlinearity in the DETF sensing element is amplitude-stiffening (or 'hard-spring effect'). This nonlinearity with the stiffening-spring variety will result in the resonant frequency increasing at large vibration amplitudes of the DETF tines. Since the MEMS RXL measures input acceleration through the change of resonant frequency of DETF sensing element, this amplitude-frequency conversion induced by the mechanical nonlinearity will directly influence the performance of RXL. Therefore, the amplitude of driving voltage (v_d) in the frequency-tracking oscillator must be controlled to limit the oscillation amplitude of the DETF tines. The nonlinear stiffening-spring variety and the limit of amplitude, a_c, (where the amplitude-frequency relationship of DETF sensing element becomes nonlinear), are discussed using the nonlinear model introduced below.

To model the stiffening-spring nonlinear effect, a cubic spring term, $k_{m3}q_1^3$, is added to Equation (6.6) in the case of fundamental resonant mode ($i=1$):

$$\left(\int_0^{L_T} \rho \cdot t_T \cdot w_T \phi_1^2 dx + \int_0^{L_T} m_{Ele} \cdot \phi_1 \left(L_T/2 \right) \cdot \delta \left(L_T/2 \right) \cdot \phi_i dx \right) \ddot{q}_1$$

$$+ \left(\int_0^{L_T} \left(E \cdot \frac{t_T \cdot w_T^3}{12} \frac{\partial^4 \phi_1}{\partial x^4} + F_{Axial} \frac{\partial^2 \phi_1}{\partial x^2} \right) \cdot \phi_1 dx \right) q_1 + k_{m3}q_1^3 = 0$$

$$(6.39)$$

Introduce Equation (6.37) in the case of fundamental resonant mode (*i=1*) into Equation (6.39):

$$\ddot{q}_1 + \omega_1^2 q_1 + \frac{k_{m3}}{M_{eff}} q_1^3 = 0 \tag{6.40}$$

The solution of $q_1(t)$ is a periodic function that has the frequency depending on the amplitude of vibration. Assuming the frequency perturbation, $\Delta\omega$, by the stiffening-spring nonlinear effect is much smaller than the linear resonant frequency of the fundamental mode, ω_1, the $q_1(t)$ can be written as follows:

$$q_1(t) = a_0 + a_1 \cos\left[(\omega_1 + \Delta\omega)t\right] \tag{6.41}$$

Substituting Equation (6.41) into Equation (6.39):

$$- a_1 (\omega_1 + \Delta\omega)^2 \cos\left[(\omega_1 + \Delta\omega)\,t\right] + a_0 \omega_1^2 + a_1 \omega_1^2 \cos\left[(\omega_1 + \Delta\omega)\,t\right]$$

$$+ \frac{k_{m3}}{M_{eff}} a_0^3 + 2\frac{k_{m3}}{M_{eff}} a_0^2 a_1 \cos\left[(\omega_1 + \Delta\omega)\,t\right] + 2\frac{k_{m3}}{M_{eff}} a_0 a_1^2 \cos^2$$

$$[(\omega_1 + \Delta\omega)\,t] + \frac{k_{m3}}{M_{eff}} a_1^3 \cos^3\left[(\omega_1 + \Delta\omega)\,t\right] = 0 \tag{6.42}$$

In order to make Equation (6.42) true, the constant terms must be equal to zero:

$$a_0 \omega_1^2 + a_0 \frac{k_{m3}}{M_{eff}} = 0$$

$$\Rightarrow a_0 = 0 \tag{6.43}$$

Equation (6.42) can be simplified to:

$$- a_1 (\omega_1 + \Delta\omega)^2 \cos\left[(\omega_1 + \Delta\omega)\,t\right] + a_1 \omega_1^2 \cos\left[(\omega_1 + \Delta\omega)\,t\right]$$

$$+ \frac{k_{m3}}{M_{eff}} a_1^3 \cos^3\left[(\omega_1 + \Delta\omega)\,t\right] = 0 \tag{6.44}$$

Applying the transformation of trigonometric function on the cubic term in Equation (6.44):

$$- a_1 (\omega_1 + \Delta\omega)^2 \cos\left[(\omega_1 + \Delta\omega)\,t\right] + a_1 \omega_1^2 \cos\left[(\omega_1 + \Delta\omega)\,t\right]$$

$$+ \frac{k_{m3}}{M_{eff}} \frac{a_1^3}{2} \left\{ \frac{3}{2} \cos\left[(\omega_1 + \Delta\omega)\,t\right] + \frac{1}{2} \cos\left[(\omega_1 + \Delta\omega)\,t\right] \right\} = 0 \tag{6.45}$$

Since only the fundamental terms are of interest, the higher-order terms can be ignored in the following analysis. Letting the sum of fundamental terms equal to zero, the expression of frequency perturbation can be obtained as:

$$\Delta\omega \approx \frac{3k_{m3}}{8\omega_1 M_{eff}}a_1^2, \text{ when } \Delta\omega \ll \omega_1 \tag{6.46}$$

The nonlinear stiffness k_{m3} can be evaluated from the elastic potential energy of the tine of DETF and is given by [12]:

$$k_{m3} = \frac{1}{2}\frac{Ew_T t_T}{L_T}\left[\int_0^{L_T}\left(\frac{d\phi_1}{dx}\right)^2 dx\right]^2 \tag{6.47}$$

where $\phi_1 x$ is the fundamental mode shape of the DETF sensing element. According to Equations (6.46) and (6.47), the nonlinear stiffness, k_{m3}, is always positive and the resonant frequency of DETF sensing element may increase with the increase in the oscillation amplitude of the tines.

When the DETF sensing element is embedded in the frequency-tracking oscillator, it will be driven by an external harmonic force and will also have damping. To evaluate the response of the DETF sensing element with a nonlinear stiffness coefficient under this condition, Equation (6.39) will be solved in a forced, damped regime:

$$M_{eff}\ddot{q}_1 + 2\lambda\dot{q}_1 + \eta q_1^2\dot{q}_1 + \omega_1^2 M_{eff}q_1 + k_{m3}q_1^3 = F_{Act}\cos(\omega t) \tag{6.48}$$

where λ is the linear damping ratio and η is the coeffcient of nonlinear damping- the damping that increases with the amplitude of oscillation. $F_{Act}\cos\omega t$ is the harmonic actuation force applied on the DETF sensing element. Equation (6.61) is also known as "Duffing Equation" in the particular case of h = 0, which has been well studied. There are various methods to solve the "Duffing Equation" [22] and it has been proved that when the actuation force exceeds a certain limit (see Figure 6.26), the solution of $q_1(\omega)$ will deviate from the linear resonant response (dot line in Figure 6.26) and will have two saddle-node bifurcation points. A linear stability analysis can confirm that the upper and lower branches (solid line in Figure 6.26) of the response are stable solutions and the middle branch (dash line in Figure 6.26) that exists between the two saddle-nodes is unstable.

The nonlinear frequency-amplitude relationship shown in Figure 6.26 may result in hysteretic response of the frequency-tracking oscillator or make the oscillator randomly switch in the bifurcation region, in which case the

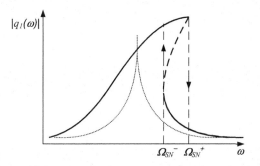

Figure 6.26 Nonlinear frequency-amplitude relationship.

output signal frequency of oscillator might not be able to precisely track the resonant frequency of the DETF sensing element that changes with the input acceleration. Therefore, finding the nonlinear limit of the magnitude of the actuation force will provide a useful guidance for the oscillator circuit design.

When the magnitude of the actuation force is reduced, two saddle-node bifurcation points will move close to each other and merge into an inflection point for a critical magnitude of actuation force. This critical magnitude of actuation force ($|F_{Act}|_c$) and the corresponding limit of oscillation amplitude of DETF tines ($|a|_c$) can be solved using secular perturbation theory from Equation (6.48) [22]:

$$|F_{Act}|_c^2 = \frac{32}{27} \frac{9 + \tilde{\eta}^2}{\left(\sqrt{3} - \tilde{\eta}\right)^3} \cdot (2\lambda)^3 \cdot \frac{\omega_1^3}{k_{m3}} \tag{6.49}$$

$$a_c^2 = \frac{8}{3} \frac{1}{\sqrt{3} - \tilde{\eta}} \cdot \frac{2\lambda}{M_{eff} k_{m3}} \tag{6.50}$$

where $\tilde{\eta} = \dfrac{\eta \omega_1}{k_{m3}}$.

By applying the fundamental mode shape of the DETF sensing element into Equation (6.47), assuming the linear damping ration $2\lambda = \omega_1/Q$ (Q is the quality factor of DETF sensing element in linear resonant) and ignoring the nonlinear damping coefficient, the critical oscillation amplitude can be correlated to the geometry dimension of the DETF sensing element as:

$$a_c \approx C \cdot \frac{w_T}{\sqrt{Q}} \tag{6.51}$$

Depending on the different mode shape approximations used in the derivation, the coefficient C varies from 1.209 [23] to 1.463 [24]. For the DETF sensing element designed in Section 6.2.6 and assuming a quality factor equal to 10000, its critical oscillation amplitude will be about 0.036–0.044 μm. However, on the other hand, the small oscillation amplitude may cause difficulty for the oscillator front-end amplifier to pick up the motional current signal of the DETF sensing element from the noise floor and degrade the resolution of the RXL.

- **Nonlinearity of Electro-Mechanical Coupling**

Another dominant nonlinearity in the operation of the DETF sensing element is due to the nonlinear actuator capacitance that will induce in an actuation force inversely proportional to the square of displacement of resonator beam. In order to understand the effect of nonlinear electro-mechanical coupling on the DETF resonant sensor, we need to rewrite the Equation (6.48):

$$\frac{1}{2}\left(\frac{\partial C(q_1)}{\partial q_1}\right)(V_p - v_d)^2 = M_{eff}\ddot{q}_1 + 2\lambda\dot{q}_1 + \eta q_1^2\dot{q}_1 + \omega_1^2 M_{eff}q_1 + k_{m3}q_1^3$$

$$(6.52)$$

The derivative of $C(q_1)$ is not a constant but a function of transverse displacement of the tine, q_1. Replace the derivative of $C(x)$ with a series expansion around the equilibrium position ($q_1 = 0$), Equation (6.52) becomes,

$$\frac{1}{2}\left(\frac{\partial C(0)}{\partial q_1} + \frac{1}{1!}\frac{\partial^2 C(0)}{\partial q_1^2}q + \frac{1}{2!}\frac{\partial^3 C(0)}{\partial q_1^3}q_1^2 + \cdots\right)(V_P - v_d)^2$$
$$= M_{eff}\ddot{q}_1 + 2\lambda\dot{q}_1 + \eta q_1^2\dot{q}_1 + \omega_1^2 M_{eff}q_1 + k_{m3}q_1^3 \qquad (6.53)$$

On the left hand side of Equation (6.53), the terms with even powers of q_1 will contribute to the static force and the force in phase with the drive voltage, v_d at the fundamental frequency, ω_1, whereas the terms with odd powers of q_1 will contribute to the force that is in phase with the position of the beam, q_1, and thus responsible for the nonlinear actuation force. As this force is a function of q_1, its behavior is similar to a non-linear spring which is named as electrostatic spring here. The above equation can be linearized if we assume the damping of the system is very low which is true for the DETF sensing element. In a weak damping system, it only responses to the input which is close to its natural frequency, ω_1, and the Equation (6.53) can be simplified as:

$$\frac{\partial C(0)}{\partial q_1}V_p\,|v_d|\cos(\omega t) = M_{eff}\ddot{q}_1 + 2\lambda\dot{q}_1$$
$$+ \eta q_1^2\dot{q}_1 + (K_{eff} - k_{e1})q_1 + (k_{m3} - k_{e3})q_1^3$$

$$(6.54)$$

And the electrostatic spring coefficients are given by:

$$k_{e1} = \left(\frac{V_p^2}{2} + \frac{|v_d|^2}{8} \right) \frac{\partial^2 C(0)}{\partial q_1^2} \qquad (6.55)$$

$$k_{e3} = \left(\frac{V_p^2}{16} + \frac{|v_d|^2}{96} \right) \frac{\partial^4 C(0)}{\partial q_1^4} \qquad (6.56)$$

Similar to the analysis for the mechanical nonlinear model, the resonant frequency of DETF sensing element including the electro-mechanical coupling nonlinear effect can be estimated as:

$$\omega_1'^2 = \frac{k_{eff} - k_{e1}}{M_{eff}} + \frac{3}{4} \frac{k_{m3} - k_{e3}}{M_{eff}} |q_1|^2 \qquad (6.57)$$

As the electrostatic spring coefficients always tending to reduce the effective stiffness of the system, this nonlinear effect is called electrostatic spring softening effect. Moreover, if the value of the third-order electrostatic spring coefficient, k_{e3}, is equal to the mechanical stiffening-spring coefficient, k_{m3}, the Duffing term will be cancelled out from Equation (6.52). This implies that the mechanical nonlinearity of the DETF sensing element might be cancelled on the first-order approximation by a specific electro-mechanical transducer design combining with certain magnitudes of a DC bias voltage and an AC driving voltage.

After considering the mechanical and electro-mechanical coupling nonlinear effects on the DETF sensing element, the equivalent electrical circuit model in Figure 6.25 is modified as shown in Figure 6.27:

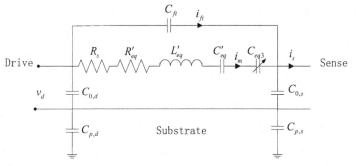

Figure 6.27 Equivalent electrical circuit for DETF with parasitic effect and nonlinear effects.

The equivalent electrical components for this model and above equation are as follows:

$$L'_{eq} = \frac{M_{eff}}{2K_P K_I} \tag{6.58}$$

$$C'_{eq} = \frac{2K_P K_I}{K_{eff} - k_{e1}} \tag{6.59}$$

$$R'_{eq} = \frac{\sqrt{M_{eff}(K_{eff} - k_{e1})}}{2QK_P K_I} \tag{6.60}$$

$$C_{eq3} = \frac{32}{3} \frac{(K_{eff} - k_{e1}) K_P^2 K_I^2}{M_{eff}(k_{m3} - k_{e3})} \cdot \frac{1}{|i_m|^2} \tag{6.61}$$

Comparing to the model shown in Figure 6.25, an additional variable capacitance, C_{eq3}, is added in the equivalent electrical circuit model to represent the influence of cubic spring term on the response of DETF sensing element. Also, the first-order electrostatic spring coefficient, k_{e1}, is included in the expressions of the equivalent capacitance, C'_{eq}, and resistance, R'_{eq}, to represent its influence on the resonant frequency of DETF sensing element.

6.3.3 Design of Frequency Tracking Oscillator

To monitor the resonant frequency variation of the DETF sensing element with input acceleration, an oscillator circuit is needed to sustain the oscillating vibration of two tines of DETF within certain frequency range and translate the mechanical vibration to electronic voltage signal. A schematic view of an oscillator circuit is shown in Figure 6.28.

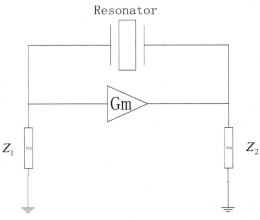

Figure 6.28   Schematic view of an Electro-mechanical oscillator circuit.

As a positive feedback circuit, the criterion for oscillation at angular frequency ω_0 can be summarized as amplitude condition [25]:

$$|G(j\omega_0)| = \left| \frac{G_m Z_1 Z_2}{Z_1 + Z_2 + Z_r} \right| \geq 1 \tag{6.62}$$

And phase condition:

$$\angle G(j\omega_0) = \arctan \frac{\text{Im}(G(j\omega_0))}{\text{Re}(G(j\omega_0))} = 0 \tag{6.63}$$

where Z_1 and Z_2 are the input/output impendence of the DETF sensing element, G_m is the gain of the amplifier and other feedback components.

In order to satisfy the amplitude condition of oscillating, an amplifier is necessary in the oscillator which compensates the energy loss from the mechanical vibration of DETF tines to maintain the loop-gain always equal to or greater than one. For the parallel-plate electro-mechanical transducer working with DETF sensing element, the driving signal is voltage whereas the sensing single is current. So, a trans-impedance amplifier will be connected to the sensing port to transfer the current to voltage. On one side, the gain of amplifier should be large enough to initiate the oscillation from random electronic noise and sustain the oscillation by compensating any electronic or mechanical loss in the oscillator. On the other hand, in order to avoid overdriving the DETF sensing element, which may result in a nonlinear response of the DETF and degrade the resolution of RXL, an amplitude control mechanism that limits the peak actuation voltage needs to be added before the driving port. Typically, a variable gain amplifier (VGA) or a clip-clip amplifier is used in the circuit design. Ideally, all of the signal circulating in the oscillator will pass through the DETF sensing element. However, because of the feedthrough capacitance (see Figure 6.27), C_{ft}, between the driving and sensing ports of the DETF sensing element, there will be unwanted feedthrough current directly coupled into the sensing current of the DETF through the C_{ft}. This parasitic current will attenuate or even blanket out the current signal from the DETF. It can be proven that the feedthrough capacitance cannot exceed [21]:

$$Max(C_{ft}) = \frac{Q}{2} C_{eq} = \frac{1}{4\pi f_c R_{eq}} \tag{6.64}$$

or the oscillation may not be possible.

In order to satisfy the phase condition, the sum of phase shift on each part of circuit should be equal to 0 or $2n\pi$. Ideally, the phase shift on the DETF

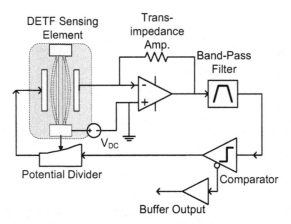

Figure 6.29   Topology of square-wave electro-mechanical oscillator.

sensing element is 0 or π, which depends on polarity of DC voltage applied to the driving and sensing ports, and the phase shift of remaining electronic circuit is designed to be near zero or p at the resonant frequency of the DETF. However, Because of the presence of feedthrough capacitance, an extra phase shift will be added in the oscillator circuit as well, which needs to be cancelled by adding phase correction. Some other parasitic effects from the bonding wires, electronic elements, chip socket and print circuit board (PCB) may also introduce additional phase shift which needs to be compensated as well.

One possible oscillator circuitry topology is shown in Figure 6.29, where the trans-impedance amplifier is used as a front-end stage to amplify and convert the weak motion current signal of DETF into voltage, band-pass filter is used for phase compensation, comparator and potential divider are used for actuation level control [21, 26].

6.3.4 Conclusion

This chapter introduces the design methods for particular topologies of MEMS RXLs with DETF sensing elements and micro-lever amplifiers. Both mechanical design and electronic design of the RXL are studied based on analytic models as well as FEM simulations. Several design examples of the RXL are shown and evaluated using FEM simulations. Depending on the application requirements, the design of RXL may emphasise different aspects. The design of the RXL may also need iterative modifications to adapt certain manufacturing processes and achieve reasonable yield.

6.4 Seismic Acceleration Resolution

The MEMS RXL measures input acceleration by measuring the shift in the resonant frequency of the DETF sensing element using a frequency-tracking oscillator. Assuming that the DETF sensing element is operated in the linear regime and the resonant frequency is only changed by input acceleration and the frequency of the oscillator output signal is exactly equal to the resonant frequency of DETF sensing element, then the resolution of MEMS RXL is determined by the resolution of the frequency counter used to record the data. A resolution of 1 mHz for an averaging time of 1 s for a signal frequency of around 100 KHz can be readily achieved, e.g. using Keysight 53230A. For an accelerometer with a scale factor of 100 Hz/m/s^2 this measuring capability is corresponding to an acceleration resolution of 10^{-8}m/s^2/Hz$^{1/2}$. However, the true resolution of the MEMS resonant accelerometer can be limited by other factors that are discussed in the following sections.

6.4.1 Frequency Noise Model

Since the resonant frequency of the DETF sensing element might be influenced by many other inherent and external factors, and the frequency of oscillator output signal might not precisely equal the resonant frequency of the DETF but limited by electronic noise. The resolution of MEMS RXL is often limited by the frequency fluctuations of the output signal of the frequency tracking oscillator. The minimum detectable acceleration of MEMS RXL is defined as:

$$a_{\min} = \left(\frac{\Delta f_n}{f_0} \right) \cdot \frac{\bar{f}_0}{S_{Axl}} \tag{6.65}$$

where the $(\Delta f_n / f_0)$ is the fractional frequency resolution of the output signal of the frequency tracking oscillator with invariant acceleration input, normally represented in the unit of ppm (part per million) or ppb (part per billion), $\bar{f}_0$ is the average frequency of the oscillator output signal over a certain period of measurement and S_{Axl} is the scale factor of accelerometer. Therefore, the resolution of the MEMS RXL can be studied and evaluated from the frequency stability of the oscillator output signal.

The ideal output of the oscillator in the MEMS RXL should be a pure sinusoid signal whose phase (φ) is a constant and the angular frequency (ω_0)

is only a function of input acceleration:

$$v(t) = V_0 \sin\left[\omega_0(a_{in})t + \varphi\right] \tag{6.66}$$

However, the real output signal of an oscillator fluctuates in amplitude and phase even with invariant input acceleration. The fluctuations in amplitude and phase can be characterised by amplitude and phase noise, respectively (see Figure 6.30).

Assuming "narrow band noise", which means the time scale of the amplitude/phase fluctuation induced by the noise is much longer than the reciprocal of oscillator frequency, the output signal of oscillator can be represented as [27]:

$$v(t) = [1 + \alpha(t)]\, V_0 \sin\left[\omega_0 t + \varphi(t)\right], |\alpha(t)| \ll 1, |\varphi(t)| \ll 1 \tag{6.67}$$

Amplitude noise is not as critical to the resolution of MEMS RXL because counting techniques provide good immunity to amplitude noise. However, the phase noise will directly impact the frequency of the output signal measured by the counter, resulting in frequency fluctuations seen in the recorded data.

In order to understand the influence of phase noise on the output signal, it is useful to review the theory of phase modulation first. A phase modulated

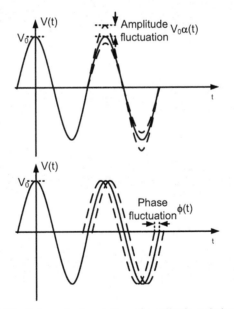

Figure 6.30   Schematic description of amplitude and phase noises.

sinusoid wave can be written as:

$$v(t) = V_0 \sin[\omega_0 t + \theta(t)] \tag{6.68}$$

In the simplest case, the phase modulation term is set as:

$$\theta(t) = \theta \sin(pt) \tag{6.69}$$

Substituting the above equation into Equation (6.68) and expanding:

$$v(t) = V_0[\sin \omega_0 t \cos(\theta \sin pt) + \cos \omega_0 t \sin(\theta \sin pt)] \tag{6.70}$$

Using the identities:

$$\cos(x \sin \phi) \equiv J_0(x) + 2[J_2(x) \cos 2\phi + J_4(x) \cos 4\phi + \cdots]$$
$$\sin(x \sin \phi) \equiv 2[J_1 \sin \phi + J_3(x) \sin 3\phi + \cdots] \tag{6.71}$$

where $J_0(x), J_1(x), \cdots$ are the Bessel Functions of x and order 0, 1..., respectively. The Equation (6.69) is rewritten as:

$$v(t) = V_0 [J_0(\theta) \sin \omega_0 t + J_1(\theta) \sin(\omega_0 + p)t - J_1(\theta) \sin(\omega_0 - p)t$$
$$+ J_2(\theta) \sin(\omega_0 + 2p)t + J_2(\theta) \sin(\omega_0 - 2p)t + \cdots] \tag{6.72}$$

The nth order of Bessel Functions is identified as:

$$J_n(x) \equiv \left(\frac{x}{2}\right)^n \left[\frac{1}{n!} - \left(\frac{x}{2}\right)^2 \frac{1}{1!(n+1)!} + \left(\frac{x}{4}\right)^4 \frac{1}{2!(n+2)!} - \cdots\right] \tag{6.73}$$

Thus, for a small value of θ (reasonable for the low-noise assumption mentioned above), the approximations of Bessel Function is:

$$J_0(\theta) \approx 1$$
$$J_1(\theta) \approx \frac{\theta}{2}$$
$$J_2(\theta) \approx J_3(\theta) \cdots \approx 0$$

And the phase modulated output signal is given by:

$$v(t) = V_0 \left[\sin \omega_0 t + \frac{\theta}{2} \sin(\omega_0 + p)t - \frac{\theta}{2} \sin(\omega_0 - p)t\right] \tag{6.74}$$

The influence of phase noise on the frequency of oscillator output signal can be derived from Equation (6.74):

$$\tilde{v}(t) = V_0 \left[\sin \omega_0 t + \frac{\varphi}{2} \sin(\omega_0 + \tilde{p})t - \frac{\varphi}{2} \sin(\omega_0 - \tilde{p})t\right] \tag{6.75}$$

where the phase noise ($\varphi(t) = \varphi \sin(\tilde{p}t)$) can be seen as a random phase modulation signal. It is clear that the phase noise will generate frequency fluctuations in oscillator output signal and degrade the resolution of the MEMS RXL. In the time domain, this random phase modulation signal has random amplitude and frequency. But in the frequency domain, the power of $\varphi(t)$ may have a certain distribution depending on the different frequencies [28], which can be represented in the form of PSD.

There are two useful tools for characterizing the frequency fluctuations of oscillator: one is the phase/frequency noise power spectral density (PSD) which is a useful description for the frequency-domain noise performance of oscillators and the other is the Allan Variance (AVAR) which provides an established time-domain metric for oscillator frequency stability [29].

6.4.1.1 PSD of phase/frequency noise

From the view of spectrum, a pure sinusoid signal will exhibit an infinite power at its frequency point or a delta function shape spectrum. With noise, the spectrum of the signal will be broadened. In general cases, the noise power associated with the most interesting noise phenomena is clustered in a very narrow band around the "carrier frequency" (this name borrowed from the communication technology and it means the ideal resonant frequency of DETF sensing element here). One common PSD description for the phase noise is $S_\varphi(f)$ which is defined as the one-sided power spectral density of the random phase fluctuation $\varphi(t)$. The physical dimension of $S_\varphi(f)$ is rad^2/Hz.

As described by Equation (6.75), a sinusoid signal polluted by the phase noise will exhibit random frequency fluctuations. Frequency is the time derivative of phase. Hence, the power spectral density of frequency noise is given by:

$$S_{\Delta v}(f) = f^2 S_\varphi(f) \tag{6.76}$$

In frequency metrology, the fractional power spectral density of frequency noise is more common:

$$S_y(f) = \frac{f^2}{v_0^2} S_\varphi(f) \tag{6.77}$$

Note that f designates the Fourier frequency in the spectral analysis and v_0 designates the carrier frequency. The minimum detectable input acceleration, or in other words, the resolution of MEMS RXL can be expressed as:

$$a_{\min}^2(B) = \left(\frac{\bar{v}_0^2}{S_{Axl}^2}\right) \int_{1/\tau d}^{B} S_y(f) df \tag{6.78}$$

Table 6.5 Most common encountered phase-noise processes

Law	Slope	Noise Processes	Units of b_i
$b_0 f^0$	0	White phase noise	rad^2/Hz
$b_{-1} f^{-1}$	-1	Flicker phase noise	rad^2
$b_{-2} f^{-2}$	-2	White frequency noise	$rad^2 Hz$
$b_{-3} f^{-3}$	-3	Flicker frequency noise	$rad^2 Hz^2$
$b_{-4} f^{-4}$	-4	Random walk of frequency	$rad^2 Hz^3$

where B is the bandwidth of the frequency tracking oscillator output signal and τ_d is the temporal scale limit when the frequency drift phenomenon impacts the output signal.

According to Equation (6.78), the minimum detectable input acceleration of the accelerometer is determined by the measurement bandwidth and the functional form of the power spectral density of phase/frequency noise. The bandwidth is related to the range of frequencies over which the input acceleration is to be measured. Roughly speaking, a large bandwidth is necessary to measure the input acceleration over a wide frequency range. To describe and study the PSD of phase/frequency noise, an indispensable model is the power-law function [30]:

$$S_\varphi(f) = \sum_{\substack{i=-4 \\ (or\ less)}}^{0} b_i f^i \tag{6.79}$$

Normally, the spectra of the phase noise are plotted on a log-log scale; hence, a term f^i will map to a straight line of slope i. The most commonly encountered phase-noise processes and their power-law characterization is given in Table 6.5 [30].

6.4.1.2 Allan variance

Another commonly used metric for the time-domain characterization of oscillators is the Allan Variance (AVAR). The Allan variance is defined as the expectation of the two-sample variance [29]:

$$\sigma_y^2(\tau) = \mathbf{E} \left\{ \frac{1}{2} [\bar{y}_{k+1} - \bar{y}_k]^2 \right\} \tag{6.80}$$

In practice, the simple mean replaces the statistical expectation. If N contiguous samples $\bar{y}_k(\tau)$ are measured (τ is the measurement time of each sample),

Table 6.6 Noise types, Power spectral densities and Allan variance

Noise Type	$S_\varphi(f)$	$S_y(f)$	$S_\varphi(f) \leftrightarrow S_y(f)$	$\sigma_y^2(\tau)^*$
White phase noise	b_0	$h_2 f^2$	$h_2 = \dfrac{b_0}{v_0^2}$	$\dfrac{3 f_H h_2}{(2\pi)^2}\tau^{-2}$
Flicker phase noise	$b_{-1}f^{-1}$	$h_1 f$	$h_1 = \dfrac{b_{-1}}{v_0^2}$	$[1.038 + 3\ln(2\pi f_H \tau)] \times$ $\dfrac{h_1}{(2\pi)^2}\tau^{-2}$
White frequency noise	$b_{-2}f^{-2}$	h_0	$h_0 = \dfrac{b_{-2}}{v_0^2}$	$\dfrac{1}{2}h_0\tau^{-1}$
Flicker frequency noise	$b_{-3}f^{-3}$	$h_{-1}f^{-1}$	$h_{-1} = \dfrac{b_{-3}}{v_0^2}$	$2\ln(2)h_{-1}$
Random walk frequency noise	$b_{-4}f^{-4}$	$h_{-2}f^{-2}$	$h_{-2} = \dfrac{b_{-4}}{v_0^2}$	$\dfrac{(2\pi)^2}{6}h_{-2}\tau$

$^*f_H$ is the low-pass cut-off frequency, needed for the noise power to remain finite.

then (N–1) differences $\bar{y}_{k+1} - \bar{y}_k$ will be calculated and the measured Allan variance is:

$$\sigma_y^2(\tau) = \frac{1}{2(N-1)} \sum_{k=1}^{N-1} (\bar{y}_{k+1} - \bar{y}_k)^2 \tag{6.81}$$

Table 6.6 [30] is a summary of the relationship between the phase and frequency power density spectra and the Allan variance. The relationship between $S_\varphi(f)$ and $S_y(f)$ is exact, but the conversion between power density spectra and variance is always approximate as the frequency integrals [31] of the conversion cannot be evaluated completely.

For the MEMS RXL, the square root of AVAR can be related to the minimum detectable input acceleration or the resolution of the accelerometer. It can be seen from Table 6.6 that the Allan variance can be reduced by increasing measurement time, τ, only for the white and flicker phase noise and the white frequency noise, the Allan variance is a constant independent of measurement time for flicker frequency noise, and for random walk or even high order frequency noise, the Allan variance will increase with the measurement time. Moreover, the measurement time is also inversely proportion to the highest measurable frequency of the input acceleration. Therefore, the measurement time will be based on the noise behaviour as well as the measurement bandwidth requirement of the MEMS resonant accelerometer.

6.4.2 Factors Influencing Resolution

In this section, several factors underlying the resolution of MEMS RXLs are studied. As discussed before, the resolution of MEMS RXLs studied here is determined by the frequency stability of the output signal of the frequency tracking oscillator. It is shown that the noise originating from the DETF sensing element and electronic components in the frequency tracking oscillator are translated into the output signal of the oscillator by the feedback loop. Also, the mechanical and electrostatic nonlinearity in the DETF sensing element and the parallel-plate electro-mechanical transducers causes a non-linear mixing effect which may directly convert the voltage noise of the DC bias and AC driving signal into frequency fluctuations of the output signal.

6.4.2.1 Phase noise of the DETF sensing element

The noise in the DETF resonator can be classified as dissipation-induced noise and parametric noise. Systems that dissipate energy are necessarily sources of noise as per the fluctuation-dissipation theorem [32]. This theorem can also be applied to mechanical resonators with nonzero dissipation, and so the DETF sensing element will also be a source of noise. Since the energy dissipated in the oscillation tines is mainly transformed to heat, the dissipation-induced noise of the mechanical resonator is also called mechanical-thermal noise. For parametric noise, there are various sources, such as local temperature fluctuations, absorption-desorption processes and nonlinear effects of the DETF sensing element.

● **Mechanical-Thermal Noise**

The dissipation-induced noise equally partitions its energy into phase noise and amplitude noise. Assuming the DETF sensing element is driven within the linear range and ignoring the nonlinear mixing effect of the parallel-plate transducer, then only the phase noise will be interested in the analysis of the accelerometer resolution. The power spectral density of phase noise at frequency f from the carrier frequency is given by [32]:

$$S_{\varphi\_thermal}(f) = \frac{v_0}{(2v_0 f + f^2)^2 + (v_0^2/Q)^2} \frac{k_B T}{8\pi^3 |x_0|^2 M_{eff} Q} \qquad (6.82)$$

where ν_0 is the resonant frequency of the DETF sensing element, Q is the quality factor, k_B is the Boltzmann constant, T is the temperature, $|x_0|$ is the maximum deflection of the DETF beam in vibration, and M_{eff} is the effective mass of one DETF tine. If we define the energy E_c of the vibration beam at the resonant frequency and the power P_c needed to maintain the oscillation amplitude:

$$E_c = \frac{1}{2} M_{eff} (2\pi \nu_0)^2 |x_0| \tag{6.83}$$

$$P_c = \nu_0 E_c / Q \tag{6.84}$$

and consider the frequency offset that are well off the peak resonance, $f \gg \nu_0/2Q$, but small compared to the carrier frequency, $f \ll \nu_0$, the $S_\varphi(f)$ can be approximated as:

$$S_{\varphi\_thermal}(f) \approx \frac{k_B T}{4 P_c Q^2} \left(\frac{\nu_0}{f} \right)^2 \tag{6.85}$$

The above equation indicates that at a frequency that is far from carrier frequency, the phase noise induced by mechanical-thermal noise is a random walk phase noise or white frequency noise (see Table 6.7). The corresponding fractional power spectral density of frequency noise is:

$$S_{y\_thermal}(f) \approx \frac{k_B T}{4 P_c Q^2} \tag{6.86}$$

According to Equations (6.82) and (6.84), the PSD of thermal-mechanical phase noise is inversely proportional to the square of oscillation amplitude, $|x_0|$, and the quality factor, Q, of the DETF beam. So increasing the magnitude of driving signal and quality factor will be helpful to lower the thermal-mechanical phase noise. However, if the amplitude of resonant motion exceeds a critical value (see Equation (6.63)), it may result in mechanical nonlinearity of DETF sensing element as discussed in Section 6.3.2. By utilizing the spring-softening effect of the electrostatic nonlinearity, this amplitude limit of linear resonant motion might be increased to further reduce the thermal-mechanical phase noise [33].

• **Parametric Noise**

The temperature fluctuation and adsorption-desorption process of the resonator will induce parametric phase noise at the output signal of the DETF sensing element. [34, 35] The name "parametric" comes from the generation mechanisms of noise: For temperature fluctuations, the effects on the resonator behaviour include changing the density and elastic modulus to directly affecting the resonant frequency and through changing the dimensions and energy dissipation. For adsorption-desorption processes, the nonzero pressure of surface-contaminating molecules adsorb to or desorb from the resonator surface, changing the effective mass of the resonator and thereby changing its resonance frequency. These two noise mechanisms have significant effects on sub-micron or nano-scale resonators because of their very small thermal capacity and mass [35, 36]. However, for the DETF sensing elements in the MEMS RXL with beam dimensions in several tens to hundreds micrometres, their influence on the phase noise behaviour of the DETF can be neglected. However, as we will discuss in Section 4, the temperature has remarkable effects on the mid to long term frequency stability of the resonator.

• **Nonlinear Noise Mixing**

The phase noise induced by nonlinear effects originates from two mechanisms. One mechanism originates from the presence of the first order electrostatic stiffness coefficient (k_{e1}). According to the model in Figure 2.42 and the expression of k_{e1}, the resonant frequency of DETF sensing element will be modulated by the DC bias voltage and the magnitude of AC driving signal as:

$$\omega_{1e}^2 = \frac{K_{\!\mathit{eff}} - \left(\frac{V_P^2}{2} + \frac{|v_d|^2}{8}\right)\frac{\partial C(0)}{\partial q_1^2}}{M_{\!\mathit{eff}}} \tag{6.87}$$

Equation (6.86) indicates that if there is any 'low-frequency' noise (the frequency band of noise is much lower than the resonant frequency of the DETF) interfering with the DC bias voltage or the magnitude of AC driving signal, it will be converted to frequency fluctuations of the DETF sensing element.

The other mechanism originates from the presence of the cubic stiffness coefficient ($k_{m3} - k_{e3}$). As shown in Equation (6.70), the fluctuations of the oscillation amplitude will change the resonant frequency proportionally to the cubic stiffness coefficient. This mechanism will not only convert the amplitude noise generated by energy dissipation or other physical process to the phase/frequency noise of the DETF, but also convert the voltage noise of AC driving signal as well as DC bias voltage into the phase/frequency noise of the DETF as discussed below:

By assuming that the DETF sensing element is driven within a quasi-linear range and the quality factor of the DETF is high enough, the oscillation amplitude of DETF can be related to the AC driving signal and DC bias [37]:

$$|x_0| = \left(\frac{2Q|v_d|}{M_{eff}\omega_{1e}^2} \right) K_P \tag{6.88}$$

where ω_{e1} is defined by Equation (6.87) and K_P is defined by Equation (6.51).

As the oscillation amplitude of the DETF is proportional to the DC bias voltage and magnitude of the AC driving signal, the resonant frequency of DETF sensing element can be expressed as:

$$\omega_1'^2(|v_d|, V_P) = \omega_{1e}^2 + \frac{3}{4} \frac{k_{m3} - k_{e3}\,(|v_d|, V_P)}{M_{eff}} \left(\frac{2Q|v_d|}{M_{eff}\,\omega_{1e}^2} \right)^2 K_P^2(V_P) \tag{6.89}$$

According to Equation (6.88), because of the direct amplitude-frequency nonlinear conversion, the 'low-frequency' noise (the frequency band of noise is much lower than the resonant frequency of DETF) varying the DC bias voltage and magnitude of the AC driving signal will be translated into the frequency fluctuation of DETF sensing element, even though the cubic stiffness coefficient ($k_{m3} - k_{e3}$) might be cancelled by specific combination of DC bias voltage and magnitude of the AC driving signal.

The low-frequency voltage noise may come from the source of DC bias voltage, the electronic components of frequency tracking oscillator or other environmental fluctuations and it is critical to limit the low-frequency voltage noise for improving the resolution of the MEMS RXL. Because of the nonlinear noise conversion mechanisms in the DETF sensing element, the phase/frequency noise level of the DETF can be directly related to the voltage noise level of DC bias voltage as well as the AC driving signal. Moreover, because of the voltage-frequency conversion relationship, if the PSD of voltage noise shows a $1/f^N$ dependence of frequency, the PSD of voltage-converted phase noise of the DETF sensing element output signal will show a $1/f^{N+2}$ dependence of offset frequency from the carrier. That means that voltage-noise up-converted phase noise might significantly influence the resolution of MEMS RXL, particularly for low frequencies acceleration signal measurement.

6.4.2.2 Noise in semiconductor amplifiers

Electronic noise exists in all the electronic components in the frequency-tracking oscillator circuit. The noise in an amplifier can be classified as additive noise and parametric noise. If a noise, either a voltage or current noise source, which can be added to the signal, it will be referred to as additive noise. If a noise can be presented by modulating the carrier in amplitude, in phase or in both with a near-dc process, it will be named as parametric noise. Additive noise originates in the region around the carrier frequency, while parametric noise comes from near-dc [30]. The most significant difference of these two types of noises is that additive noise is always present, while the parametric noise requires the presence of the nonlinear effect and a carrier signal. The PSD of additive noise is generally not related to the Fourier frequency, but the PSD of parametric noise may vary with the Fourier frequency away from the carrier frequency.

White and flicker noises are additive and parametric noises in semiconductor amplifiers, respectively. The effect of both white noise and flicker noises in the amplifier will be discussed below.

• **White Noise**

All noise processes with white-spectrum statistics are classified under the label of additive noise. The noise is not correlated with the signal. In this case, the noise power is equally partitioned into two degrees of freedom, half in amplitude noise and half in phase noise. Two mainly sources of white noise in the electronic device are the electronic-thermal noise and shot noise. The equations of them are given below:

$$V_{\text{thermal noise}} = \sqrt{4k_B T R B} \tag{6.90}$$

$$I_{\text{shot noise}} = \sqrt{2q I_{dc} B} \tag{6.91}$$

where B is the measurement bandwidth, R is the resistance, q is the electron charge, and I_{dc} is the average DC current. And the PSD of the phase noise contributed by the white noise is given by [38]:

$$S_{\varphi\_white}(f) = b_0, b_0 = \frac{NB}{P_0} \tag{6.92}$$

where NB is the noise power in the bandwidth B and P_0 is the power of carrier signal.

• **Flicker Noise**

Being different from the white noise, flicker noise has a power spectrum scaling of the type $1/f^x, x \in (0.8, 1.2)$ [39]. As shown in Figure 3.2, since the flicker noise only shows up when the carrier signal present at the amplifier output is large enough and the flicker noise sidebands grow in proportion

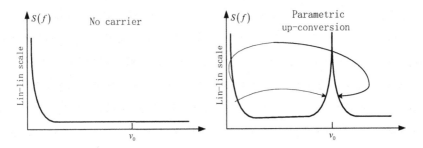

Figure 6.31   Parametric up-conversion of near-dc flicker in amplifiers.

to the carrier power [38], it is obvious that the close-in flicker noise results from the near-dc flicker noise modulated by the carrier through certain up-conversion mechanisms (so it is classified as parametric noise). Two simple models are discussed below:

The first mechanism of the noise up-conversion can contribute to the amplifier's non-linearity. The output of the amplifier can be represented by a polynomial and truncated at the second degree:

$$v_0(t) = a_1 v_i(t) + a_2 v_i^2(t) + \cdots \tag{6.93}$$

The coefficient a_1 is the linear voltage gain and the a_2 is the second order factor. The noisy input signal of the amplifier is given by:

$$v_i(t) = V_i e^{j\omega_0 t} + n(t) = V_i e^{j\omega_0 t} + n'(t) + jn''(t) \tag{6.94}$$

where $V_i e^{j\omega_0 t}$ is the carrier signal, $n(t) = n'(t) + jn''(t)$ is the near-DC noise which may be carried by the carrier or internally generated by the amplifier. Combining the above two equations and ignoring the terms far from the carrier frequency, the output signal is given by:

$$v_0(t) = V_i \left\{ a_1 + 2a_2 \left[n'(t) + jn''(t) \right] \right\} e^{j\omega_0 t} \tag{6.95}$$

In the above equation, the term $V_i a_1 e^{j\omega_0 t}$ is the output carrier and the term $V_i 2a_2 \left[n'(t) + jn''(t) \right] e^{j\omega_0 t}$ is the close-in noise added on the output carrier signal. The random amplitude and phase fluctuation components are:

$$\alpha(t) = 2\frac{a_2}{a_1} n'(t) \tag{6.96}$$

$$\varphi(t) = 2\frac{a_2}{a_1} n''(t) \tag{6.97}$$

Since the random amplitude and phase fluctuation shown in Equations (6.92) and (6.93) are independent of the carrier signal amplitude, the power spectral densities $S_\alpha(f)$ and $S_\varphi(f)$ are expected to be independent of the carrier power.

Another mechanism of near-dc noise up-conversion needs to be introduced is the fluctuation of the amplifier gain. This effect can be modelled by a random time-varied voltage gain instead of a constant value:

$$A(t) = A_0\left[1 + e(t)\right] = A_0\left[1 + e'(t) + e''(t)\right] \qquad (6.98)$$

where $e(t) = e'(t) + e''(t)$ is the amplifier gain fluctuation caused by near-dc noise. For instance, in a bipolar transistor, $e''(t)$ is easily identified as a signal proportional to the fluctuating voltage which modulates the collector-base reverse voltage and the capacitance between them. The change of collector-base capacitance will affect the amplifier phase lag and generate phase noise. By applying an identical carrier signal at the input of the amplifier, the output signal will be:

$$v_0(t) = A_0\left[1 + e'(t) + je''(t)\right]V_i e^{j\omega_0 t} \qquad (6.99)$$

And the random amplitude and phase fluctuation components are:

$$\alpha(t) = e'(t) \qquad (6.100)$$
$$\varphi(t) = e''(t) \qquad (6.101)$$

Similar to the first mechanism, $\alpha(t)$ and $\varphi(t)$ are independent of the input amplitude and hence $S_\alpha(f)$ and $S_\varphi(f)$ are expected to be independent of the carrier power as well. It must also point out that the second parametric model is nearly linear in contrast with the first model based on strong nonlinearity. In real amplifiers, both mechanisms may exist and generate flicker noise.

The amplitude noise of amplifiers may influence the resolution of MEMS RXL because of the nonlinear effect present in the DETF sensing element as discussed before. In that case, the amplitude noise contributed by amplifiers could influence the magnitude of the AC driving signal, which may result in the fluctuation of the resonant frequency of the DETF sensing element. Comparing to the amplitude noise, the phase noise contributed by all amplifiers is more important for the resolution analysis of MEMS RXL, which may directly increase the phase/frequency noise of the output signal of the frequency tracking oscillator. If it can be assumed that the white and flicker noise are independent, the PSD of phase noise, $S_{\varphi\_a}(f)$, caused by the amplifier noise is expressed as:

$$S_{\varphi\_amp}(f) = b_0 + b_{-1}\frac{1}{f}$$

$$b_0 = \frac{NB}{P_0}$$

$$b_{-1} = const. \tag{6.102}$$

To reduce the phase noise contributed by the amplifier, low-noise amplifier with good gain linearity should be selected for the frequency tracking oscillator. A stable, low-noise supply voltage for the amplifier is important to reduce the phase noise of amplifier as well. Moreover, since the phase lag and gain of an amplifier are sensitive to temperature, a fluctuating environment might generate amplitude and phase noise by the parametric modulation process. It is observed [30] that the temperature fluctuations will induce a noise spectrum proportional to $1/f^5$ at a very small frequency offset from the carrier frequency.

6.4.2.3 Noise in the frequency tracking oscillator

In this sub-section, two noise phenomena in the frequency-tracking oscillator will be discussed. One phenomenon is the 'Leeson effect', by which the noise of the sustaining amplifiers in the oscillator circuitry is turned into frequency noise in the output of the oscillator. The other phenomenon is that the oscillator interacts with the phase/frequency noise of the DETF sensing element and translate it to the output of the oscillator.

The Leeson effect

In order to study the effect of feedback loop on the amplifier-induced phase noise, or called Leeson effect [40], a simplified oscillator noise model is given by Figure 6.32, where the amplifier has ideal gain A, the transfer function, $\beta(j\omega)$, represents the feedback effects of the DETF sensing element and other electronic components, and the random phase term, $e^{j\psi}$, represents the phase noise caused by the amplifiers only. All other noises being ignored in this model.

Firstly, for the slow time-varying phase component, $\psi_L(t)$, of the random phase term, if assuming that the phase fluctuation of $\psi_L(t)$ is slower than the inverse of the relaxation time of the DETF sensing element, $\tau = Q/\pi\nu_0$, which means the DETF sensing element can effectively feed the phase

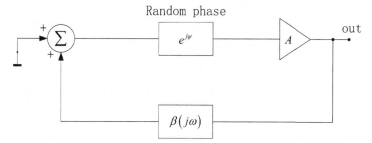

Figure 6.32 Simplified oscillator noise model for Leeson effect analysis.

fluctuation back to the input of the loop, then phase $\psi_L(t)$ will be treated as a quasi-static perturbation. In that case, the oscillator will respond to this phase perturbation with a frequency fluctuation:

$$(\Delta v)(t) = \frac{v_0}{2Q}\psi_L(t) \qquad (6.103)$$

And the corresponding power spectral density is:

$$S_{\Delta v} = \left(\frac{v_0}{2Q}\right)^2 S_{\psi_L}(f) \qquad (6.104)$$

The instantaneous output phase is:

$$\varphi(t) = 2\pi \int (\Delta v)(t)dt \qquad (6.105)$$

Since the time-domain integration associates to a multiplication by $1/j\omega = 1/j2\pi f$ in the Fourier transform, the slow phase-fluctuation spectrum is:

$$S_\varphi(f) = \frac{1}{f^2}\left(\frac{v_0}{2Q}\right)^2 S_{\psi_L}(f) \qquad (6.106)$$

Assuming that the phase fluctuation of $\psi_H(t)$ is faster than the inverse of the DETF sensing element relaxation time, loosely speaking, the resonator will not respond to these fast phase fluctuations and thereby, no phase fluctuation will enter the loop but the phase fluctuation will directly go through the amplifier and show up in the output signal of oscillator. In this case, no noise regeneration takes place in the feedback loop, thus:

$$\varphi'(t) = \psi_H(t) \qquad (6.107)$$
$$S'_\varphi(t) = S_{\psi_H}(t) \qquad (6.108)$$

By adding the fast and slow phase-fluctuation spectrum, the Leeson's formula that represents the relations between the oscillator output phase spectrums to the amplifier phase fluctuation is expressed as: [40]

$$S_\varphi(f) = \left[1 + \frac{1}{f^2}\left(\frac{v_0}{2Q}\right)^2\right] S_\psi(f) \quad \text{(Leeson formula)} \quad (6.109)$$

To rewrite the Leeson's formula as:

$$S_\varphi(f) = \left(1 + \frac{f_L^2}{f^2}\right) S_\psi(f) \quad (6.110)$$

where

$$f_L = \frac{v_0}{2Q} \quad (6.111)$$

is the Leeson frequency. This formula was proposed by David B. Leeson as a model for short-term frequency fluctuations and was initially intended to explain the phase noise of the crystal-oscillator used for airborne Doppler radar. Because of the Leeson effect, the shape of the amplifier phase noise spectrum whose offset frequencies are smaller than the Leeson frequency will become steeper where the order of slope is increased by two in log-log scaling maps, whereas the shape of the other parts of the noise spectrum which are far from the Leeson frequency will not change, which increases the power density of the close-in phase noise of the oscillator output signal. The re-shaped amplifier phase noise may further reduce the short-term frequency stability of oscillator output signal, and correspondingly reduce the resolution of the MEMS RXL.

Interaction with DETF sensing element

The interaction between the electronic circuit and DETF sensing element can be divided into two aspects. The first is the voltage-frequency noise conversion effect. As discussed in Section 6.3.2, the resonant frequency of the DETF sensing element can be modulated by the DC bias voltage and the magnitude of the AC driving signal, the modulation equation is rewritten here:

$$\omega_1'^2(|v_d|, V_P) = \frac{K_{\mathit{eff}}}{M_{\mathit{eff}}} - \frac{k_{e1}(|v_d|, V_P)}{M_{\mathit{eff}}} + \frac{3}{4}\frac{k_{m3} - k_{e3}(|v_d|, V_P)}{M_{\mathit{eff}}}$$
$$\left(\frac{2Q|v_d|}{M_{\mathit{eff}}\omega_1^2}\right)^2\left(\frac{\varepsilon L_{Elc}t_T}{g_{Elc}^2}V_P\right)^2 \quad (6.112)$$

Define a fluctuating DC bias voltage,

$$V_P(t) = V_{P0} + \Delta V_P(t) \tag{6.113}$$

and a fluctuating magnitude of AC driving signal:

$$|v_d|(t) = |v_{d0}| + |\Delta v_d|(t) \tag{6.114}$$

If assuming that the voltage fluctuation is significantly slower than the resonant frequency of DETF, which means that the DETF sensing element can effectively feed the frequency fluctuation back to the input of the loop, the frequency fluctuation will be treated as a quasi-static perturbation. By assuming that the fluctuation of the DC bias voltage is independent to the fluctuation of AC driving signal, the frequency fluctuation induced by the magnitude of DC bias voltage is expressed as:

$$\Delta \omega_{1\_DC}^{'2}(t) \approx \frac{1}{4\omega_1^{'2}(V_{P0},|v_{d0}|)} \left\{ \frac{2}{M_{eff}} \frac{\varepsilon t L_{elc}}{g_{elc}^3} V_{P0} + \frac{3}{4M_{eff}} \right.$$

$$\left. \left(\frac{V_{P0}^3}{8} + \frac{V_{P0}|v_{d0}|^2}{48} \right) \left(\frac{2Q|v_{d0}|}{M_{eff}\omega_1^2} \right)^2 \left(4\frac{\varepsilon^3 L_{Elc}^3 t_T^3}{g_{Elc}^9} \right) \right\}^2 \Delta V_P^2(t)$$

$$= G_{DC} \cdot \Delta V_P^2(t), \tag{6.115}$$

where $|\Delta \omega_{1\_DC}'(t)| \ll |\omega_1'(V_{P0},|v_{d0}|)|$ and $|\Delta V_P(t)| \ll |V_{P0}|$

And the frequency fluctuation induced by the magnitude of AC driving signal is expressed as:

$$\Delta \omega_{1\_AC}^{'2}(t) \approx \frac{1}{4\omega_1^{'2}(V_{P0},|v_{d0}|)} \left\{ \frac{2}{M_{eff}} \frac{\varepsilon t L_{elc}}{g_{elc}^3} V_{P0} + \frac{3}{4M_{eff}} \right.$$

$$\left. \left(\frac{V_{P0}^2|V_{do}|}{8} + \frac{|v_{d0}|^3}{48} \right) \left(\frac{2Q}{M_{eff}\omega_1^2} \right)^2 \left(4\frac{\varepsilon^3 L_{Elc}^3 t_T^3}{g_{Elc}^9} V_{P0}^2 \right) \right\}^2 |\Delta V_d|^2(t)$$

$$= G_{AC} \cdot |\Delta V_d|^2(t),$$

where $|\Delta \omega_{1\_AC}'(t)| \ll |\omega_1'(V_{P0},|v_{d0}|)|$ and $|\Delta V_d|(t)| \ll |V_{d0}|$ (6.116)

where the variables in Equations (6.115) and (6.116) are consistent with those defined for the modelling in Section 6.3.

Setting the noise spectrum of DC bias voltage as $S_{DC}(f)$ and the amplitude noise spectrum of the AC driving signal is $S_{\alpha\_AC}(f)$, then the phase-fluctuation spectrum of the DETF sensing element induced by the nonlinear voltage-frequency conversion effect is:

$$S_{\varphi\_v-f}(f) = \frac{1}{f^2} \left[G_{DC} \cdot S_{DC}(f) + G_{AC} \cdot S_{\alpha\_Ac}(f) \right] \qquad (6.117)$$

The second aspect of the interaction between the electronic circuit and the DETF sensing element is the reshaping of the open-loop phase noise profile of the DETF sensing element in the feedback loop. Similar to the amplifier phase noise, the close-in phase noise of DETF sensing element will also be re-shaped because of the Leeson effect. The phase noise PSD of the DETF sensing element may show $1/f^3$ and $1/f^4$ dependence when the Fourier frequency is close to the frequency of carrier signal, which are referred to as flicker frequency noise and random walk frequency noise, respectively. At very low offset frequency from the carrier, the $1/f^4$ noise is the dominant process before frequency and aging impact drift.

6.4.3 Estimation of Resonant Seismic Sensors' Resolution

In this sub-section, the resolution of a single-axis MEMS RXL will be estimated based on the analyses in the previous sections. The mechanical design of this MEMS RXL is summarised in Tables 6.3 and 6.4.

6.4.3.1 Mechanical-thermal noise limited resolution

In this part, only the influence of the phase noise generated from the dissipation-fluctuation mechanism of the DETF resonator, which is the fundamental limit on the resolution of MEMS RXL, is considered.

First, the scale factor of the DETF sensing element is rewritten here:

$$\frac{\delta\nu}{\nu_c} \approx \frac{FS}{4} \qquad (6.118)$$

$$S = 0.293 \left(\frac{L_T^2}{Et_T w_T^3} \right) \qquad (6.119)$$

where the δf and f_c in Equation (6.9) are replaced by $\delta\nu$ and ν_c here in order to distinguish it from the Fourier frequency, f, in the following derivation.

Reproducing Equation (6.82), the phase noise density spectrum is:

$$S_{\varphi\_thermal}(f) = \frac{\nu_c}{(2\nu_c f + f^2)^2 + (\nu_c^2/Q)^2} \frac{k_B T}{8\pi^3 |x_0|^2 M_{eff} Q} \qquad (6.120)$$

And the fractional frequency noise density spectrum is given by:

$$S_{y\_thermal}(f) = \left(\frac{f^2}{\nu_c^2}\right) S_{\varphi\_thermal}(f)$$

$$= \frac{f^2}{\nu_c \left(2\nu_c f + f^2\right)^2 + \left(\nu_c^5/Q^2\right)} \frac{k_B T}{8\pi^3 |x_0|^2 M_{eff} Q} \tag{6.121}$$

According to Equation (6.78), the mechanical-thermal noise limited resolution of MEMS RXL is:

$$a_{min\_thermal}^2(B) = \left(\frac{\bar{\nu_c^2}}{S_{Axl}^2}\right) \int_{1/\tau d}^{B} S_{y\_thermal}(f) df \tag{6.122}$$

Since the resolution is a function of measurement bandwidth, Equation (6.120) is calculated for different bandwidth values: **B** = 1–500 Hz by using MATLAB and the results are shown in Figure 6.33. The parameter values used in above calculation are listed in Table 6.8.

As discussed before, in thea case that the offset frequency, f, satisfies the inequality $\nu_c/Q \ll f \ll \nu_c$, the mechanical-thermal noise can be approximated to a white frequency noise:

$$S'_{y\_thermal}(f) \approx \frac{k_B T}{4 P_c Q^2} \tag{6.123}$$

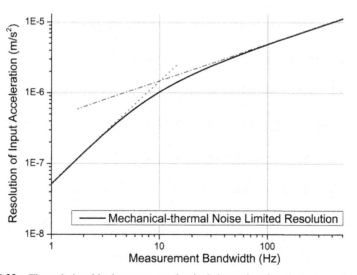

Figure 6.33 The relationship between mechanical-thermal noise limited resolution and measurement bandwidth.

Table 6.7 Summaries of design parameters of MEMS RXL for noise-limited resolution analysis

Parameter Name	Value
ν_c	111.49 kHz
Q	10000
$|x_0|$	0.1 μm
k_B	1.38×10^{-23} J/K
M_{eff}	70.99 ng
T	300 K
S_{Axl}	143.5 Hz/m/s^2

And Equation (6.120) can be simplified as:

$$a_{min\_thermal}^2(B) = \left(\frac{\bar{\nu}_c^2}{S_{Axl}^2} \right) \int_{1/\tau d}^{B} S_{y\_thermal}(f) df$$

$$\approx \left(\frac{\bar{\nu}_c^2}{S_{Axl}^2} \right) \left[C_1 + S'_{y\_thermal} \cdot \left(B - 10 \frac{\nu_c}{Q} \right) \right] \quad (6.124)$$

For the accelerometer example shown in Table 6.7, the $\nu_c/Q \approx 11\,Hz$, and therefore the resolution of input acceleration is proportion to $B^{1/2}$ (see the red dash-dot line in Figure 6.33) when bandwidth is larger than 100 Hz. In the case that the offset frequency, f, is smaller than ν_c/Q, the mechanical-thermal noise can be approximated to a white phase noise:

$$S''_{y\_thermal}(f) \approx \frac{f^2 Q}{\nu_c^5} \cdot \frac{k_B T}{8\pi^3 |x_0|^2 M_{eff}} \quad (6.125)$$

and the resolution of input acceleration is therefore proportional to $B^{3/2}$(see the blue dot line in Figure 6.33) when the bandwidth is smaller than 10 Hz.

However, when the DETF sensing element is operated in the frequency-tracking oscillator, the close-in phase noise will be reshaped because of the 'Leeson Effect' and Equation (6.119) is modified to:

$$a'^2_{min\_thermal}(B) = \left(\frac{\bar{\nu}_c^2}{S_{Axl}^2} \right) \int_{1/\tau d}^{B} \left(1 + \frac{f_L^2}{f^2} \right) S_{y\_thermal}(f) df \quad (6.126)$$

where $f_L = \nu_c/2Q$. Recalculating the resolution from Equation (6.126) (results shown in Figure 6.34) and it can be noticed that the resolution for low measurement bandwidth is 10 times worse after considered Leeson Effect. Moreover, the resolution of input acceleration is now proportional to the $B^{1/2}$

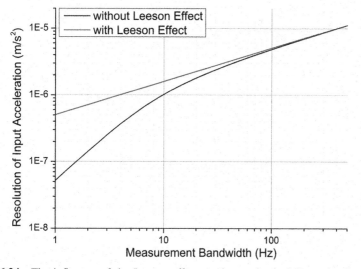

Figure 6.34 The influence of the Leeson effect on the mechanical thermal noise limited resolution.

for all measurement bandwidth because of the Leeson Effect, which indicates that the mechanical-thermal noise of the DETF sensing element can always be approximated to a white frequency noise when the DETF forms an oscillator.

6.4.3.2 Electronic noise-limited resolution

As discussed before, the influence of electronic noises on the resolution of MEMS RXL results from the noises in the amplifiers and the voltage-frequency noise conversion effect.

The phase noise arising from the noises in the amplifiers is given by:

$$S_{\varphi\_amp}(f) = \frac{Fk_BT}{P_0} + \sum_i b_{-1i}\frac{1}{f} \tag{6.127}$$

Since the Leeson Effect also needs to be included in the estimation of resolution, the amplifier phase noise limited resolution of MEMS RXL is expressed as:

$$a_{min\_amp}^2(B) = \left(\frac{\bar{\nu_c^2}}{S_{Axl}^2}\right) \int_{1/\tau d}^{B} \left(1 + \frac{f_L^2}{f^2}\right)\left(\frac{f^2}{\nu_c^2}\right) S_{\varphi\_amp}(f)df \tag{6.128}$$

The power of AC driving signal is set to be -30 dBm (10^{-6} W) and the frequency tracking oscillator is assumed to consist of three cascading

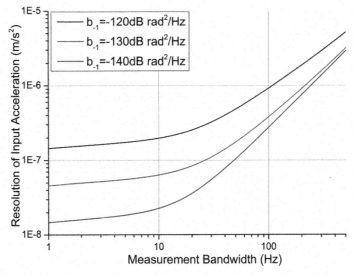

Figure 6.35 Calculated amplifier noise limited resolution versus the measurement bandwidth.

amplifier with equal flicker phase noise coefficient, b_{-1}. Depending on the amplifiers employed, b_{-1} may vary from -120 to -140 dB rad^2/Hz [38]. All other parameters used in the calculation are still same as those in Table 6.7 and the results are shown in Figure 6.35.

Since the close-in phase noise of amplifiers is reshaped by the Leeson effect, flicker phase noise becomes flicker frequency noise which limits the resolution for narrow measurement bandwidths below 10 Hz. Therefore, the amplifier with low flicker phase noise is very critical for the design of MEMS RXLs.

Another impact of electronic noise on the resolution of MEMS RXLs is the frequency noise induced by voltage-frequency noise conversion effect. In the frequency tracking oscillator of MEMS RXLs, the driving signal is filtered by a comparator before being fed back in to the DETF sensing element. Therefore, the amplitude noise of the AC driving signal can be ignored and only the noise in DC bias voltage needs to be included in the following analysis.

The PSD of voltage noise in DC bias voltage can be modelled as:

$$S_{DC}(f) = b_{0,Dc} + \frac{b_{-1,DC}}{f} + \frac{b_{-2,DC}}{f^2} \qquad (6.129)$$

Table 6.8 Summaries of noise models for regulated DC source and battery

	Agilent E3631A	Agilent B2962A	Battery [41]
$b_{0,DC}$ (V$^2_{rms}$/Hz)	4.9×10^{-17}	1×10^{-18}	1.6×10^{-19}
$b_{-1,DC}$(V$^2_{rms}$/Hz·s)	2.05×10^{-14}	2.05×10^{-15}	2×10^{-17}
$b_{-2,DC}$(V$^2_{rms}$/Hz·s^2)	1×10^{-13}	2.5×10^{-14}	1×10^{-14}

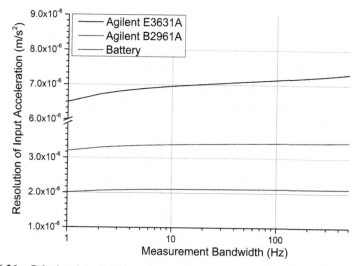

Figure 6.36 Calculated the DC bias voltage noise limited resolution versus the measurement bandwidth.

The DC bias voltage for the prototype resonant accelerometer may come from a DC power supply or battery. Table 6.8 gives the PSD coefficients of two DC power supplies and a battery.

Then the DC bias voltage noise limited resolution of MEMS RXL can be estimated by:

$$a^2_{min\_DC}(B) = \left(\frac{\bar{\nu}^2_c}{S^2_{Axl}}\right) \int^B_{1/\tau d} \left(\frac{1}{\nu^2_c}\right) G_{DC}S_{DC}(f)df \qquad (6.130)$$

The calculated DC bias voltage noise limited resolution is shown in Figure 6.36. The parameters used in the calculation are same as those in Table 6.7 with additional parameters listed in Table 6.9.

The $1/f$ and $1/f^2$ voltage noise of the DC bias voltage impacts the resolution of MEMS RXLs for narrow measurement bandwidth. However, since the near-DC voltage noise characteristic of DC power supply might be sensitive to the output impedance, parasitic charge, as well as the

Table 6.9   Summary of parameters for DC bias voltage noise limited resolution analysis

Parameter Name	Value
V_{p0}	10 V
$|v_{d0}|$	10 mV
L_{elc}	170 μm
t_T	25 μm
g_{elc}	2 μm
ε	8.85×10^{-12} F/m

environmental factors, such as temperature, the above estimation may require some correction by experimental data.

6.4.3.3 Combinative resolution estimation

The final resolution of MEMS RXL will be determined by the combination of all possible limit factors, including those in the above analyses. If all noise sources are assumed to be uncorrelated with each other, the final resolution of the RXL can be expressed as:

$$a_{min}^2(B) = a_{min\_thermal}^2 + a_{min\_amp}^2 + a_{min\_DC}^2 + a_{min\_other}^2 \quad (6.131)$$

In practical applications, the resolution of MEMS RXL may be limited by the ambient seismic noise, instead of the noises inherent to the MEMS sensing element and interface circuits.

6.4.4 Conclusion

This section investigates the factors underpinning the resolution of MEMS RXL. Both frequency and time domain representations of the phase noise and frequency noise are introduced. Several major noise processes in the MEMS RXL are discussed. Finally, the resolution of a prototype MEMS RXL is estimated, which is limited by different noise processes. The results indicate that the fundamental resolution of the MEMS RXL is limited by the thermal-mechanical noise of the DETF sensing element whereas the noise of DC bias voltage is the main limiting factor underlying the resolution of the prototype MEMS RXL.

6.5 Drift in Resonant Seismic Sensors

The performance of MEMS RXLs is also significantly influenced by drift of its output. Drift is defined as the non-null change of output frequency of the RXL in the absence of input acceleration over a relatively long period of time. The drift of MEMS RXLs may limit its capability for static or ultra-low frequency acceleration measurement and the repeatability of the measurements over time. Drift can be due to several reasons, such as the charge-induced bias voltage variations on the resonator, variations in ambient temperature and pressure, surface adsorption/desorption processes and aging. In this chapter, the temperature, pressure and charge-induced drift are discussed. Since the micro-mechanical structure of MEM RXLs is fabricated using a single-crystalline silicon and packaged in vacuum, the influence of aging [42] and surface adsorption/desorption [43] can be neglected.

6.5.1 Temperature Drift

The temperature drift of the MEMS RXLs is the output dependence on the temperature. In this section, several mechanisms underlying the temperature dependence of the output frequency are studied.

6.5.1.1 Temperature-dependent elasticity

The elastic properties of materials are known to be temperature dependent. The elastic behaviour of a crystalline material is determined by the strength of the bonds between the atoms. The increase of the temperature causes the elasticity to decrease, because of the distance between the atoms increases. Although this relationship is complex and non-linear [44], a linear approximation can be made if the displacement change is very small. In general, the elasticity of a material is represented by the Young's modulus (E). The change in the Young's modulus with temperature is designated as temperature coefficient of elasticity (TCE) and the expression of temperature dependent Young's modulus under the linear approximation is given by:

$$E = E_0 \left(1 + TCE \cdot \Delta T\right) \tag{6.132}$$

Since single crystalline silicon is a crystal with cubic symmetry, the Young's modulus of which is given as a second-order tensor to represent the anisotropic characteristic of the material.

$$E_{sc-si} = \begin{vmatrix} c_{11} & c_{12} & 0 & 0 & 0 & 0 \\ c_{12} & c_{11} & 0 & 0 & 0 & 0 \\ 0 & 0 & c_{11} & 0 & 0 & 0 \\ 0 & 0 & 0 & c_{44} & 0 & 0 \\ 0 & 0 & 0 & 0 & c_{44} & 0 \\ 0 & 0 & 0 & 0 & 0 & c_{44} \end{vmatrix} \tag{6.133}$$

According to the experimental results [45], the elastic stiffness coefficients and first order TCE of n-doped (0.05 Ω·cm) single crystalline silicon, which is similar to the wafer used for the fabrication of prototype MEMS RXLs, at 25°C is given as:

The temperature dependence of the Young's modulus will impact not only the natural frequency of the DETF sensing element, but also the scale factor of the MEMS RXL since the stiffness of the suspensions and micro-levers are also sensitive to the temperature change. In order to evaluate the effect of the temperature dependence of the Young's modulus on the performance of MEMS RXL, a material model including the temperature coefficients shown in Table 6.10 is substituted into the FEM model of the MEMS RXL (as shown in Figure 6.18), and the natural frequencies of DETF sensing element and scale factor of the RXL are derived from a temperature-sweep simulation. As shown in Figures 6.37 and 6.38, both natural frequency and the scale factor exhibit a linear dependence on temperature, but the TCf is negative whereas the TCS is positive. The scatter of the data points in Figure 6.38 is believed to be due to the numerical errors of the press-stressed eigenfrequency simulation.

Now, the equivalent output acceleration drift of the MEMS RXL caused by the temperature-dependence of Young's modulus can be estimated by:

$$a_{\Delta T,E}(T) = \frac{f_0 \cdot TCf \cdot (T - T_0)}{S_{Axl,0}[1 + TCS \cdot (T - T_0)]} \qquad (6.134)$$

where temperature coefficients, TCf and TCS, can be extracted from the simulation results in Figures 6.37 and 6.38. The values of $a_{\Delta T,E}$ for $T - T_0 \in [-10, 10]°C$ can be plotted as shown in Figure 6.39. A linear fit to the plotted results reveals that the equivalent output acceleration drift due to the TCE is about -0.0254 m/s^2/°C, which is a considerable source of error for the resonant accelerometer, particularly in the applications of low-frequency acceleration measurement or operating in a temperature varying environment.

Table 6.10 Elastic stiffness and temperature coefficients of single crystalline silicon

Elastic Stiffness Coefficient	Value $\times 10^9$ Pa
c_{11}	165.64
c_{12}	63.94
c_{44}	79.51
Temperature Coefficient	First Order $\times 10^{-6}$/°C
Tc_{11}	-74.87 ± 0.99
Tc_{12}	-99.46 ± 3.5
Tc_{44}	-57.98 ± 0.17

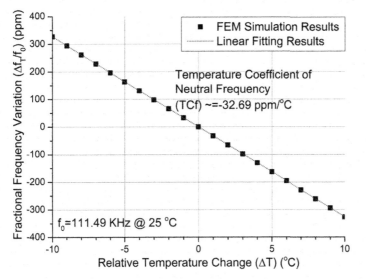

Figure 6.37 Simulated output frequency variation (due to temperature dependent elasticity) of MEMS RXL with temperature.

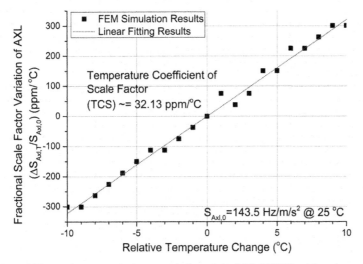

Figure 6.38 Simulated scale factor variation of the MEMS RXL with temperature.

Moreover, the TCE of single crystalline silicon may be reduced through heavy doping of single-crystal silicon [46, 47], as an alternative way to reduce the output acceleration drift-induced by the TCE.

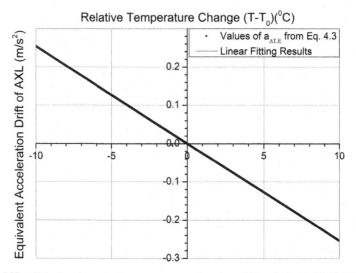

Figure 6.39   Calculated equivalent output acceleration drift of the MEMS RXL with temperature.

6.5.1.2 Thermal expansion and thermal stress

As the silicon belongs to the m3m class of the cubic crystalline materials, thermal expansion of silicon can be described by an isotropic model. So, only one coefficient of thermal expansion (CTE) of silicon is necessary. According to the reported thermal expansion measurement data of single crystalline silicon [45], the first, second and third order CTE of silicon at 25°C are summarised in Table 6.11.

Because of the thermal expansion, the dimensions of the DETF sensing element change with temperature, resulting in a change in the natural frequency of the DETF. The frequency change can result not only from the change of the dimensions of the beams, but also from the variation in the dimensions of the parallel-plate capacitive transducers. The temperature-dependent dimensions of the DETF sensing element and the

Table 6.11   First three coefficients of thermal expansion of single crystalline silicon

Coefficient of Thermal Expansion	Calculated Value at 25°C
$\alpha_{Si}^{(1)}$	$2.84 \pm 0.04 \times 10^{-6}/°C$
$\alpha_{Si}^{(2)}$	$8.5 \pm 0.5 \times 10^{-9}/°C^2$
$\alpha_{Si}^{(3)}$	$-32 \pm 2 \times 10^{-12}/°C^3$

electro-mechanical transducer may be described by a linear approximation below:

$$w_{T,\Delta T} = w_{T,0}\left(1 + \alpha_{Si}^{(1)} \cdot \Delta T\right) \tag{6.135}$$

$$L_{T,\Delta T} = L_{T,0}\left(1 + \alpha_{Si}^{(1)} \cdot \Delta T\right) \tag{6.136}$$

$$A_{Elc,\Delta T} = A_{Elc,0}\left(1 + \alpha_{Si}^{(1)} \cdot \Delta T\right)^2 \tag{6.137}$$

$$t_{Elc,\Delta T} = t_{Elc,0}\left(1 + \alpha_{Si}^{(1)} \cdot \Delta T\right) \tag{6.138}$$

$$L_{Elc,\Delta T} = L_{Elc,0}\left(1 + \alpha_{Si}^{(1)} \cdot \Delta T\right) \tag{6.139}$$

$$g_{Elc,\Delta T} = g_{Elc,0} - w_{Elc} \cdot \alpha_{Si}^{(1)} \cdot \Delta T \tag{6.140}$$

$$\rho_{Si,\Delta T} = \rho_{Si,0}\left(1 + \alpha_{Si}^{(1)} \cdot \Delta T\right)^{-3} \tag{6.141}$$

In above equations, $\alpha_{Si}^{(1)}$ is the first coefficient of thermal expansion of silicon, ΔT is the relative ambient temperature change, and the remaining variables are consistent to the model defined in Section 2.3. The subscript '0' and 'ΔT' represent the value of variables at reference and changed temperatures, respectively.

Substituting the above equations into Equation (6.70), and only considering the first order electrostatic spring-softening effect, the thermal expansion influence on the natural frequency of DETF sensing element, $f_{c,a}(\Delta T)$, can be estimated by:

$$f_{c,\alpha}^2(\Delta T) = f_{c,0}^2\left(1 + \alpha_{Si}^{(1)} \cdot \Delta T\right) - \frac{1}{4\pi^2}\left(\frac{V_P^2}{2} + \frac{|v_d|^2}{8}\right)$$
$$\frac{\varepsilon \cdot L_{Elc,0} \cdot t_{Elc,0}}{M_{eff}} \cdot \frac{\left(1 + \alpha_{Si}^{(1)} \cdot \Delta T\right)^2}{\left(g_{Elc,0} - w_{Elc} \cdot \alpha_{Si}^{(1)} \cdot \Delta T\right)^3} \tag{6.142}$$

By substituting the parameter values listed in Table 6.11 and 6.12 into Equation (6.142), the values of fractional frequency change $[f_{c,a}(\Delta T)f_{c,a}(0)]/f_{c,a}(0)$ for $\Delta T \epsilon[-10, 10]°C$ are plotted in Figure 6.40. It can be noticed that the TCf induced by the thermal expansion is much

Table 6.12 Summary of design parameters of MEMS RXL for thermal-expansion induced output drift analysis

Parameter Name	Value
V_{P0}	10 V
$\lvert v_d \rvert$	10 mV
L_{elc}	170 μm
t_T	25 μm
g_{elc}	2 μm
w_{Elc}	15 μm
ε	8.85×10^{-12} F/m

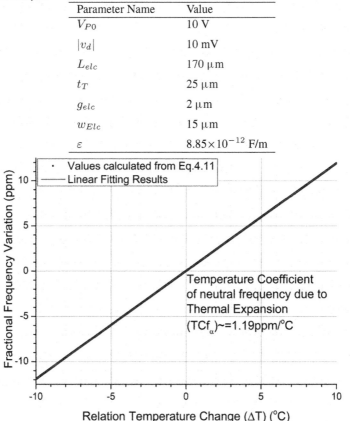

Figure 6.40 Simulated output frequency variation (due to thermal expansion) of MEMS RXL with temperature.

smaller than the TCf induced by the temperature-dependent elasticity. So, the equivalent output acceleration drift caused by the thermal expansion is also relatively small.

Furthermore, the thermal expansion of the material not only changes critical geometrical parameters, but may also result in temperature-induced stresses that could impact the output drift. This is illustrated with the prototype MEMS RXL in Figure 6.41.

With the SOI micro-machining process, a film of SiO_2 is situated between the device layer and the substrate, providing electrical isolation between the two layers. However, because of the different thermal expansions

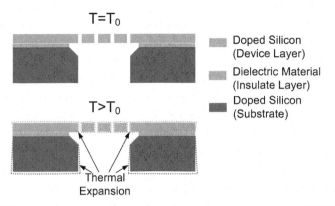

Figure 6.41   Schematic view of thermal expansion induced stress on the MEMS RXL.

of the silicon and silicon dioxide, a temperature-induced stress couples to the device through the anchors. The thermal expansion coefficient of SiO_2 is only about $5 \times 10^{-7}/°C$ [48], which is more than 50 times smaller than the thermal expansion coefficient of silicon. Since the anchors of DETF sensing element and suspensions are attached to the SiO_2 layer, the expansion of DETF and suspension towards anchors might be limited whereas the other parts, like proof-mass, can still freely expand with increasing temperature. As the DETF sensing element is connected to the proof-mass through micro-levers, the extended proof-mass will push the input end of the levers, and the extended DETF tines and connection beams of micro-levers will push the output end of the levers. Assume that the amplification factor of micro-lever is A_{Lvr}, the thermal-expansion-induced displacement at input end of the micro-lever is $T.E.D_{Lvr,in}$ and the thermal-expansion-induced displacement at output end is $T.E.D_{Lvr,out}$ for $1°C$ increase of temperature. In the case of, the DETF sensing element will experience compressive axial stress if $T.E.D_{Lvr,in} < T.E.D_{Lvr,out} \cdot A_{Lvr}$, or tensile axial stress if $T.E.D_{Lvr,in} > T.E.D_{Lvr,out} \cdot A_{Lvr}$, or no axial stress if $T.E.D_{Lvr,in} \approx T.E.D_{Lvr,out} \cdot A_{Lvr}$.

The displacement of the output end of the micro-lever can be estimated from the thermal expansion of the DETF sensing element and the connecting structures by:

$$T.E.D_{Lvr,out}(\Delta T) \approx (L_T + L_{con.}) \cdot \alpha_{Si}^{(1)} \cdot \Delta T \qquad (6.143)$$

However, the displacement estimation of the input end is dependent on the choice of topology. For instance, in the accelerometer with a differential design (the FEM model is the same as that in Figure 6.18 and the simulation results shown in Figure 6.42), as the central displacement of proof-mass is

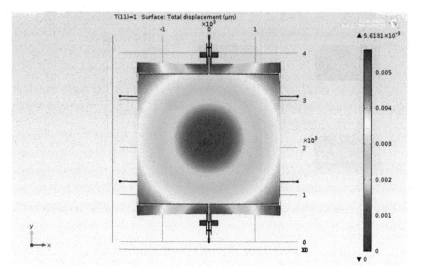

Figure 6.42   Simulated thermal expansion displacement field of single-axis RXL with differential DETFs.

zero because of the structural symmetry, the displacement of the input end can be estimated from the thermal expansion of the proof-mass:

$$T.E.D_{Lvr,in}(\Delta T) \approx \frac{1}{2}\sqrt{A_{mass}} \cdot \alpha_{Si}^{(1)} \cdot \Delta T \qquad (6.144)$$

As introduced in Section 6.2.6, the length of DETF and connection beam is about 500 mm, and the lever amplification factor of a single-axis differential accelerometer is about 50. According to the above discussion, the side length of proof-mass should be 25 mm to balance the micro-lever and cancel axial force induced on the DETF sensing element. However, since practical limitations imposed by processing considerations will limit this dimension to less than 2.5 mm, there will always be some compressive/tensile axial force induced on the DETF sensing elements when the temperature increases/ decreases.

6.5.1.3 Temperature-dependent DC bias voltage

As discussed before, the resonant frequency of the DETF sensing element can be changed by DC bias voltage because of the nonlinearity of the electro-mechanical transducer. If the DC bias voltage has a non-zero temperature coefficient, α_{T_DC}, the temperature dependent DC bias voltage applied on the DETF sensing element will result in a drift of the accelerometer output. For example, the Agilent E3631a has an output voltage temperature

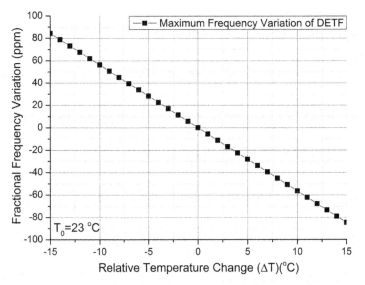

Figure 6.43 Calculated frequency variation (due to temperature-dependent DC bias voltage) of the MEMS RXL output with temperature.

coefficient equal to $\pm$ (0.01%$\times$Output+3 mV). Considering the first-order nonlinear effect of the electro-mechanical transducer, the resonant frequency of the DETF sensing element can be expressed as a function of the relative temperature change:

$$f_{T\_DC}^2 \left(\Delta T\right) = \frac{K_{eff} - \left(\dfrac{V_P^2\left(1+\alpha_{T\_DC}\Delta T\right)^2}{2} + \dfrac{|v_d|^2}{8}\right)\dfrac{\partial^2 C(0)}{\partial q_1^2}}{4\pi^2 M_{eff}} \tag{6.145}$$

Using the model parameters of DETF sensing element given by Table 6.12, the temperature dependent DC bias voltage-induced resonant frequency drift is plotted in Figure 6.43. The temperature coefficient of the natural frequency arising from the temperature dependent DC bias voltage (supplied by Agilent E3631) is about -5.7 ppm/°C.

6.5.2 Pressure-Induced Drift

MEMS resonant accelerometer is required to operate in a low pressure environment, for example, in a vacuum package. However, during the operation of the MEMS RXL, the pressure level may increase because of the leakage or the outgassing. With the increase in pressure, many factors may start impacting on the MEMS RXL, including the air damping, surface absorption/desorption processes, humidity, etc. This section will only discuss the influence of squeeze film air damping on the frequency output of the frequency tracking oscillator, caused by the increase of the pressure.

When the ambient pressure increases beyond a certain level, squeeze film air damping will arise between the moving electrode plate of the DETF sensing element and the stationary electrode plate. The squeeze film air damping force consists of two components: one is a viscous damping force; the other is an elastic force. The viscous damping force results from the viscous air flow when air is squeezed out of (or sucked into) the region between the two electrode plates. The elastic force results from the compression of air between two electrode plates. If the ambient pressure is low, the impact of squeeze film air damping can be neglected in the previous analysis for the dynamic vibration of the DETF sensing element. However, once the ambient pressure increased, the influence of squeeze film air damping needs to be considered in the oscillation equation of the DETF sensing element:

$$M_{eff} \cdot \ddot{q}_1 + (c_{vd} + c_0) \cdot \dot{q}_1 + (K_{eff} + k_{ed}) \cdot q_1 = F_{Act} \qquad (6.146)$$

where c_0 is the damping coefficient of DETF sensing element in low pressure, which is mainly due to the anchor-loss [53]and thermo-elastic damping [54], and c_{vd} is the viscous damping coefficient of the squeeze film air damping between the two parallel-plate electrodes which is given by [55]:

$$c_{vd}(\sigma) = \frac{64\sigma P_a \cdot t_{elc} \cdot L_{elc}}{\pi^6 g_{elc}}$$

$$\cdot \sum_{m,n \text{ odd}} \frac{m^2 + (n \cdot t_{elc}/L_{elc})^2}{(mn)^2 \left\{ \left[m^2 + (n \cdot t_{elc}/L_{elc})^2 \right]^2 + \sigma^2/\pi^4 \right\}} \qquad (6.147)$$

And k_{ed} is the coefficient of elastic force the squeeze film air damping which is given by [55]:

$$k_{ed}(\sigma) = \frac{64\sigma^2 P_a \cdot t_{elc} \cdot L_{elc}}{\pi^8 g_{elc}}$$

$$\cdot \sum_{m,n \text{ odd}} \frac{1}{(mn)^2 \left\{ \left[m^2 + (n \cdot t_{elc}/L_{elc})^2 \right]^2 + \sigma^2/\pi^4 \right\}} \qquad (6.148)$$

where σ is referred to as the squeeze number:

$$\sigma = \frac{12\mu_{air}\omega_c L_{elc}^2}{P_a g_{elc}^2} \left(1 + \frac{t_{elc}^2}{L_{elc}^2} \right) \qquad (6.149)$$

The P_a is the ambient pressure, μ_{air} is the coeffcient of viscosity of the air, and ω_c is the angular frequency of the DETF sensing element. The elastic force changes the output frequency of the oscillator directly through the spring constant whereas the viscous damping influences the frequency of the oscillator by the changing the damping ratio of the oscillator. So, considering the squeeze film air damping, the output frequency of the oscillator can be expressed by:

$$f_{c,damp}^2 (c_{vd}, k_{ed}) \approx \left(f_{c,0}^2 + \frac{k_{ed}}{4\pi^2 M_{eff}} \right) \left(1 + \frac{c_{vd}^2}{4\sqrt{M_{eff}K_{eff}}} \right) \qquad (6.150)$$

For a certain frequency tracking oscillator circuit, the increase of ambient pressure not only increases the output frequency but also influences the oscillator functionality. As discussed in Section 2, the oscillator must satisfy loop gain and phase conditions to lock-in to the resonant frequency of the DETF sensing element. Since both the motional resistance and the phase lag of DETF sensing element are the functions of the quality factor, the increased ambient pressure will reduce the quality factor, and the loop gain and/or phase conditions of the oscillator may be broken when the ambient pressure exceeds a certain level. So, a good vacuum packaging technique is important for the MEMS RXL, which can substantially eliminate the MEMS RXL drift resulting from the squeeze film air damping.

6.5.3 Charge-Induced Drift

In the operation of frequency-tracking oscillator, the stationary electrodes of the DETF sensing element are biased by a DC voltage which is normally an external DC source. To isolate the biased electrodes from other part of the structure, there is a layer of dielectric material between the electrodes and the substrate wafer (see Figure 6.44), for instance, the thermally grown silicon dioxide is one common dielectric material used in the MEMS fabrication process, which is called buried SiO_2(BOX). Also, there will be a thin native oxide layer over the surface of micro-machined silicon structure. Both of them could contain fixed or mobile charges [56]. Moreover, the charges may distribute in the interface between device layer and BOX or the interface between metal pad and device layer [57, 58]. The variation of charge position and/or density will directly change the effective DC bias voltage between the parallel-plate electrodes of the DETF sensing element and result in the output frequency drift of the MEMS RXL through the electrostatic spring softening effect.

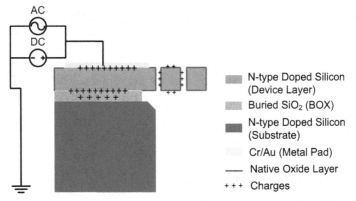

Figure 6.44   Schematic view of charge distribution in the MEMS RXL.

The charge-induced frequency drift normally appears after the power supply for the MEMS RXL is switched on. Existing charges are re-distributed by the applied electric filed and fresh charge addition takes place until the process reaches equilibrium. The time constant of this progress is dependent on the distribution of the charge, the material properties, the structure of device, as well as the external DC bias voltage. Charge drift or charge trapping phenomena has been observed in similar electrostatically-driven MEMS resonators under the influence of various DC bias voltages [56, 59]. According to reported experimental results, the observed time constant of the charge-induced surface voltage drift in MEM devices varies from several minutes to several hours. Depending on the polarity of external DC bias voltage, the output frequency drift of the MEMS RXL can be either positive or negative.

6.6 Device Fabrication and Integration

6.6.1 Micromachining Process

A foundry MEMS manufacturing process based on Silicon On Insulator (SOI) wafer, as shown in Figure 6.45, is used to fabricate the micro mechanical structure chips as described before. This process is a simple 4-mask level SOI patterning and etching micromachining process developed by MEM-SCAP.The bare chip was then glued and wire bonded to a standard LCC-44 chip carrier before test (see Figure 6.46).

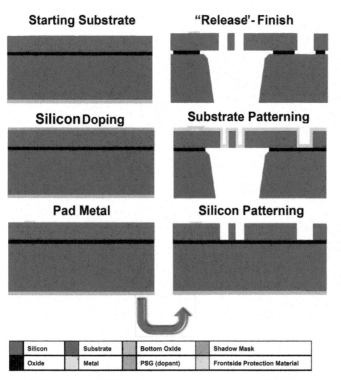

Figure 6.45 Fabrication process flow for SOIMUMPS

6.6.2 Low Pressure Package

Figure 6.47 shows the cross-section schematic view of the MEM RXL fabricated by the Silex-UOCD process. The micro-mechanical structures of MEM RXL are implemented on the device layer. In order to achieve wafer-level vacuum packaging, one side of the device wafer is fusion bonded to VIA wafer which establish vertical electrical connections from the device layer to the bond pads on the surface of the chip and the other side of the device wafer is eutectically bonded to a CAP wafer under vacuum. A layer of getter material is deposited on the inner surface of CAP wafer to adsorb the gas molecules in the cavity and maintain the vacuum level.

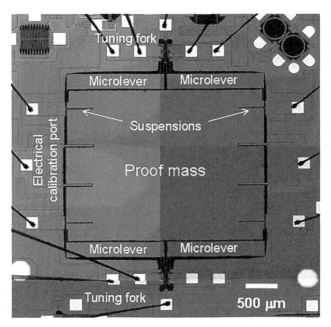

Figure 6.46   Micrograph of MEMS RXL micro-mechanical chip with bonding wires.

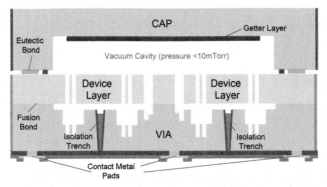

Figure 6.47   Cross-section view of MEMS RXL using Silex-UOCD fabrication process.

6.6.3 Laboratory Calibration and Results

This section introduces in-lab calibration experiments for the prototype MEMS RXLs. In order to calibrate the scale factor, resolution, dynamic range and establish the drift of the MEMS RXL, several experimental tests were carried out. Representative results from these tests are analysed and discussed in the following sections.

6.6.3.1 Experimental setup

The calibration tests include static and dynamic calibrations. The scale factor, resolution, and drift of the prototype accelerometer can be derived from the static calibration tests and the dynamic calibrations of MEMS RXL's response to real-time acceleration input excitation. Figure 6.48 shows the experimental setups for the static and dynamic calibrations. In order to attenuate the impact of the ambient seismic noise, all the calibration tests were conducted on an air-suspended vibration isolation table. The prototype MEMS RXL was clamped on a manual tilt/rotary table which was anchored on the vibration isolation table. The tilt/rotary could not only align the sensing axis of the RXL but also calibrate the scale factor of the RXL utilizing the gravity acceleration by changing the tilting angle. The electro-dynamic shaker was used to generate time/frequency varied acceleration excitation to the MEMS RXL and a macro-size CMG-3TD 3-axis seismometer was placed next to the prototype RXL to measure and record the acceleration

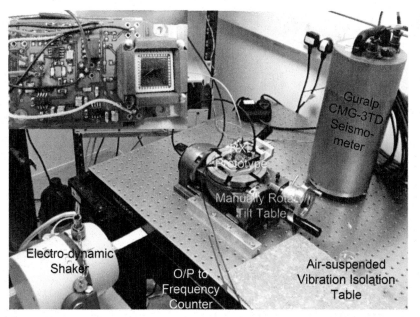

Figure 6.48 In-lab Calibration Experimental Setup (Inset: the zoomed-in photo of the prototype MEMS RXL).

excitation input simultaneously for calibration purpose. The frequency shift of the MEMS RXL was measured and recorded by the frequency counter and the data processed off-line using MATLAB.

6.6.4 Static Calibration

6.6.4.1 Accelerometer scale factor

The scale factor of RXL is derived from the output frequency shift of the RXL in a step tilting test. By changing the tilt angle of the manual table, the gravity acceleration component on the input axis of RXL will approximately change in proportional to the sinusoidal function of tilt angle. Figure 6.49 exhibits the time trace of the RXL output frequency shift during a tilt test. The angle between the RXL sensing axis and the ground was changed from 90° to 0° with 10° intervals between the steps. The output frequency of the RXL was recorded for about 300 s at each tilt position. The scale factor of RXL can be derived by correlating the average frequency shift to the equivalent gravity acceleration change on the sensing axis of RXL as plotted in Figure 6.50. The linear fitting of the experimental test results indicates that the representative RXL has a scale factor of 143.79 Hz/m/s^2, which agrees quite well with the simulation estimation (143.5 Hz/m/s^2) shown in Table 6.4.

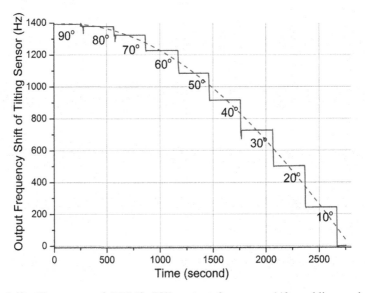

Figure 6.49 Time trace of MEMS RXL output frequency shift enabling scale factor calibration.

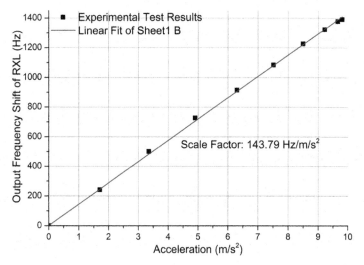

Figure 6.50   Scale factor calibration result for the prototype MEMS RXL.

However, it should be noted that because of the fabrication tolerances associated with the geometry of the design, especially the tolerances on the beam width of the DETF sensing element and suspensions and the micro-lever, the scale factor of MEMS RXL shows variations from device to device.

6.6.4.2 Accelerometer resolution

This section introduces the resolution improvements for the prototype MEMS RXL and the estimation of optimized resolution in the laboratory environment. Improving the resolution of the RXL not only requires increasing the scale factor but also optimizing the noise floor. The noise floor of a RXL may be estimated from the frequency stability of the oscillator when there is no acceleration input.

To establish the frequency stability of the sensor output, a series of measurements are made on a RXL with the sensitive axis oriented normally to the gravity field with the output logged on a frequency counter and modified Allan deviation calculations were then carried out. As a representative example, Figure 6.51 shows a series of modified Allan Deviation (the square root of MVAR) results calculated from ten separate output frequency records of a RXL with non-optimized oscillator measured by the same frequency counter over 12 hours.

The MDEV curve shown in Figure 6.51 indicates that the noise floor of the RXL is dominated by three different noise components

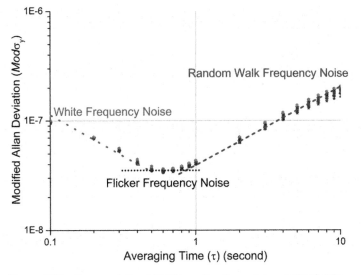

Figure 6.51 Representative MDEV results of a prototype MEMS RXL.

(see Table 6.2) – the white frequency noise, flicker (1/f) frequency noise and random walk (1/f²) frequency noise, so the resolution of RXL will be a function of the averaging time. The minimum MDEV value shown in Figure 6.51 is about 3.46×10^{-8} for a 0.6 s averaging time, corresponding to an acceleration resolution of 2.4×10^{-5} m/s². As the frequency tracking oscillator used in above MEMS RXL is not optimized, the resolution of this RXL can be further enhanced by improving the performance of the oscillator.

The frequency stability of the frequency tracking oscillator can be influenced by several factors. As discussed in Section 6.3, the front-end TIA noise and the DC bias voltage-to-frequency (V-F) noise conversion are the two potential impacting factors on the frequency stability of the oscillator. As shown in Figure 6.52, after optimizing the trans-impedance gain and using a low-noise operational amplifier, the minimum MDEV is reduced to 1.48×10^{-8} and the minimum MDEV is further reduced to 6.74×10^{-9} for 0.8 s averaging time after replacing regulated DC source with alkaline battery for DC bias, corresponding to an acceleration resolution of 4.68×10^{-6} m/s². However, it should also be noted that the output frequency of the MEMS RXL biased by batteries shows higher random walk frequency noise level and larger drift compared to the same MEMS RXL biased by a regulated DC power supply.

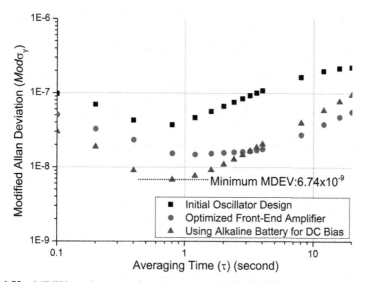

Figure 6.52 MDEV results comparison between the MEMS RXL with initial and improved oscillators.

Although in-lab calibration experiments were carried out on an air-suspended vibration isolation table, the ambient seismic noise in the laboratory can still impact the resolution of the MEMS RXL. Since the mechanical bandwidth of the RXL is designed to be at least 500 Hz, all the ambient seismic noise components below this frequency can be coupled to the DETF sensing element and degrade the frequency stability of the oscillator. Moreover, because of the low sampling rate of the frequency counter (for example, 0.1 s gate time is approximately equivalent to a sampling rate of 10 pts./second), the high frequency seismic noise components might be aliased to low frequency noise during the measurement, which will further decrease the RXL resolution for low frequency acceleration input.

To demonstrate the above analysis, seismic acceleration on the calibration platform was measured by the Guralp CMG-3TD seismometer located next to the RXL. Two different sampling rates were used in the measurement, one is 100 pts/s, which is the default sampling rate of the seismometer, and the other is 10 pts/s, which mimics the frequency counter with the gate time of 0.1 s. The mechanical bandwidth of the seismometer is 50 Hz, so the measurement using default sampling rate, 100 pts/s, is able to reproduce the real seismic acceleration profile on the platform. The measurement using the sampling rate of 10pts/s shows the frequency aliasing effect. The PSD

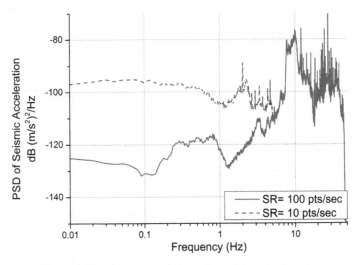

Figure 6.53 Frequency aliasing effect of the seismic noise.

of seismic acceleration calculated from the two measurements is plotted in Figure 6.53.

The red line is the PSD of seismic acceleration calculated from the measurement using default sampling rate and the blue line is the PSD of seismic acceleration calculated from the measurement using low sampling rate under nearly identical environmental conditions. It is seen from the red solid line that the seismic acceleration noise is primarily distributed above 5 Hz and the level of seismic acceleration noise is relatively low below 1 Hz, which is believed to represent the seismic noise profile of the calibration platform. As per the analysis presented above, the seismic noise level of the measurement using low sampling rate (blue dash line) is significantly higher, demonstrating the aliasing of high frequency seismic noise into low frequency content.

As introduced before, the best resolution estimation of the RXL achieved in the laboratory is about 4.68×10^{-6} m/s^2 when measurement bandwidth is 0.625 Hz (0.8 s averaging time), for a rough approximation, which is equivalent to a power spectral density of -104 dB (m/s^2)^2/Hz @ 0.625 Hz. As can be seen from the Figure 6.53, this value is comparable to the seismic noise floor plotted in blue line. So, the ambient seismic noise in the lab environment is likely to set the limit to the resolution measurement of the prototype RXL with the optimized oscillator.

6.6.5 Dynamic Calibration

This sub-section introduces the dynamic calibration of the RXL with real-time acceleration input. As shown in Figure 6.48, in the dynamic calibration, the resonant accelerometer is deployed close to a CMG-3TD seismometer on an air-suspended table. The shaker gently tapped on one side of the table at a frequency of 1 Hz to generate periodical impulse seismic signal. The real-time responses (Figure 6.54) and power spectral density analysis (Figure 6.55) of the output from the resonant accelerometer and CMG-3TD seismometer are compared respectively. As shown in Figures 6.54 and 6.55, the demodulated output signal of the prototype MEMS resonant accelerometer matches the CMG-3TD seismometer measurement result. The PSD plot shows a fundamental seismic peak at 1 Hz and harmonic peaks at 2 Hz and 4 Hz, which may originate from the 1 Hz periodical impulse seismic excitation. The sample rate cannot exceed 10 pts/s because of the limitations imposed by the counter used for recording and demodulating the frequency output.

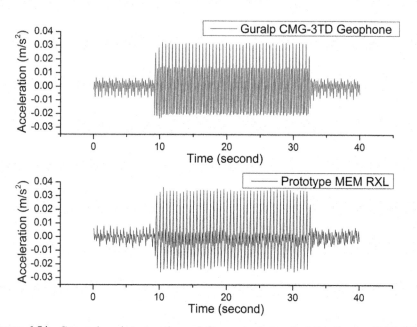

Figure 6.54 Comparison between the real-time output signal of the Guralp CMG-3TD seismometer and the demodulated output signal of the prototype MEMS RXL to the same seismic signal.

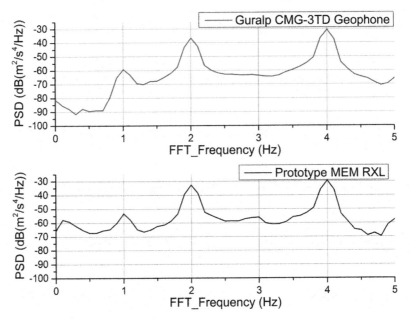

Figure 6.55   PSD analysis of the Guralp CMG-3TD seismometer and the prototype MEMS RXL response to the same real-time seismic signal.

Therefore, according to Nyquist sampling rule, seismic signals higher than 5 Hz will not be demodulated correctly.

6.6.6 Conclusion

In this sub-section, a functional prototype MEMS RXL is assembled and tested in the laboratory environment. The scale factor of the RXL was calibrated by using the gravitational load on a manual tilt/rotary table and the result shows good agreement with the FEM simulations. With an optimized frequency tracking oscillator and low noise DC bias voltage source, the minimum detectable acceleration of the prototype MEM RXL is demonstrated to be 4.68×10^{-6} m/s^2, which is comparable to the ambient seismic acceleration noise observed in the laboratory. The dynamic response of the prototype MEMS RXL is tested under a periodical excitation from a shaker and calibrated to a macro size seismometer.

References

[1] T. Aizawa, T. Kimura, T. Matsuoka, T. Takeda, and Y. Asano, "Application of MEMS accelerometer to geophysics," *International Journal of the Jcrm,* vol. 4, pp. 33–36, 2008.

[2] J. Bernstein, R. Miller, W. Kelley, and P. Ward, "Low-noise MEMS vibration sensor for geophysical applications," *Journal of Microelectromechanical Systems,* vol. 8, pp. 433–438, Dec. 1999.

[3] J. Laine, D. Mougenot, and Ieee, *Benefits of MEMS based seismic accelerometers for oil exploration,* 2007.

[4] D. J. Milligan, B. D. Homeijer, and R. G. Walmsley, "An ultra-low noise MEMS accelerometer for seismic imaging," vol. 58, pp.1281–1284, 2012.

[5] B. Homeijer, D. Lazaroff, D. Milligan, and R. Alley, "Hewlett packard's seismic grade MEMS accelerometer," in *IEEE International Conference on MICRO Electro Mechanical Systems,* pp. 585–588, 2011.

[6] U. Krishnamoorthy, R. H. O. Iii, G. R. Bogart, M. S. Baker, D. W. Carr, T. P. Swiler, et al., "In-plane MEMS-based nano-g accelerometer with sub-wavelength optical resonant sensor," *Sensors & Actuators A Physical,* vol. 145–146, pp. 283–290, 2008.

[7] K. Azgin, C. Ro, A. Torrents, and T. Akin, "A resonant tuning fork force sensor with unprecedented combination of resolution and range," in *IEEE International Conference on MICRO Electro Mechanical Systems,* pp. 545–548, 2011.

[8] E. Y. Lee, Y. Zhu, and A. A. Seshia, "A micromechanical electrometer approaching single-electron charge resolution at room temperature," in *IEEE International Conference on MICRO Electro Mechanical Systems,* pp. 948–951, 2008.

[9] R. G. Azevedo, D.G. Jones, A. V. Jog, B. Jamshidi, D. R. Myers, L. Chen, et al., "A SiC MEMS Resonant Strain Sensor for Harsh Environment Applications," *IEEE Sensors Journal,* vol. 7, pp. 568–576, 2007.

[10] J. Wang, D. Chen, L. Liu, and Z. Wu, "A micromachined resonant pressure sensor with DETFs resonator and differential structure," *in Sensors,* pp. 1321–1324, 2010.

[11] S. S. Rao, "Mechanical vibrations. 2nd ed," 2005.

[12] T. A. Roessig, "Integrated MEMS Tuning Fork Oscillators for Sensor Applications," 1998.

[13] R.E. D. Bishop, "Vibration Problems in Engineering, 4th Edition. Timoshenko S. P., Young D.H. and Weaver W. Wiley, Chichester. 1974. 538 pp. Illustrated. u00a38.95," *Aeronautical Journal,* vol. 79, pp. 138–138, 2016.

[14] E. Y. Lee, J. Yan, and A. A. Seshia, "Anchor limited Q in flexural mode resonators," *in Ultrasonics Symposium,* 2008. Ius, pp. 2213–2216, 2009.

[15] A. A. Seshia, M. Palaniapan, T. A. Roessig, and R. T. Howe, "A vacuum packaged surface micromachined resonant accelerometer," *Microelectromechanical Systems Journal of,* vol. 11, pp. 784–793, 2002.

[16] I. Zeimpekis, I. Sari, and M. Kraft, "Characterization of a Mechanical Motion Amplifier Applied to a MEMS Accelerometer," *Journal of Microelectromechanical Systems,* vol. 21, pp. 1032–1042, 2012.

[17] S. X. P. Su, H. S. Yang, and A. M. Agogino, "A resonant accelerometer with two-stage microleverage mechanisms fabricated by SOI-MEMS technology," *IEEE Sensors Journal,* vol. 5, pp. 1214–1223, 2005.

[18] X. Su, "Compliant Leverage Mechanism Design For Mems Applications," *Dissertation Abstracts International, Volume: 62-07, Section: B, page: 3357.;Chair: Alice M. Agogin,* 2001.

[19] D. Y. R. Chong, W. E. Lee, B. K. Lim, and J. H. L. Pang, "Mechanical characterization in failure strength of silicon dice," in *Thermal and Thermomechanical Phenomena in Electronic Systems, 2004. ITHERM '04. The Ninth Intersociety Conference on,* pp. 203–210, Vol. 2, 2004.

[20] T. C. Nguyen, "Micromechanical Signal Processors," *Thesis – UNIVERSITY OF CALIFORNIA,* 1994.

[21] K. E. Wojciechowski, "Electronics for Resonant Sensors."

[22] R. Lifshitz and M. C. Cross, *Nonlinear Dynamics of Nanomechanical and Micromechanical Resonators,* 2008.

[23] L. Nicu and C. Bergaud, "Modeling of a tuning fork biosensor based on the excitation of one particular resonance mode," *Journal of Micromechanics & Microengineering,* vol. 14, p. 727, 2004.

[24] J. Juillard, A. Bonnoit, E. Avignon, and S. Hentz, "From MEMS to NEMS: Closed-loop actuation of resonant beams beyond the critical Duffing amplitude," in *Sensors,* pp. 510–513, 2008.

[25] E. A. Gerber and A. Ballato, "Precision Frequency Control," 1985.

[26] E. Y. Lee, B. Bahreyni, Y. Zhu, and A. A. Seshia, "A Single-Crystal-Silicon Bulk-Acoustic-Mode Microresonator Oscillator," *IEEE Electron Device Letters,* vol. 29, pp. 701–703, 2008.

[27] W. P. Robins, "Phase noise in signal sources/Theory and applications," *London Peter Peregrinus Ltd.p,* vol. 9, 1984.

[28] A. Hajimiri and T. H. Lee, "A general theory of phase noise in electrical oscillators," *IEEE Journal of Solid-State Circuits,* vol. 42, pp. 2314–2314,1998.

[29] D. W. Allan, "Statistics of atomic frequency standards," *IEEE Proceedings,* vol. 54, pp. 221–230, 1966.

[30] E. Rubiola, "Phase noise and frequency stability in oscillators," *Cambridge University Press,* 2008.

[31] W. F. Egan, "Frequency Synthesis by Phase Lock, 2nd Edition," 1999.

[32] A. N. Cleland and M. L. Roukes, "Noise processes in nanomechanical resonators," *Journal of Applied Physics,* vol. 92, pp. 2758–2769, 2002.

[33] M. Agarwal, K. K. Park, R. N. Candler, and B. Kim, "Nonlinear Characterization of Electrostatic MEMS Resonators," in *International Frequency Control Symposium and Exposition,* pp. 209–212, 2006.

[34] J. R. Vig and Y. Kim, "Noise in microelectromechanical system resonators," *IEEE Transactions on Ultrasonics Ferroelectrics & Frequency Control,* vol. 46, pp. 1558–1565, 1999.

[35] Z. Djurić, O.Jakšić, and D. Randjelović, "Adsorption–desorption noise in micromechanical resonant structures," *Sensors & Actuators A Physical,* vol. 96, pp. 244–251, 2002.

[36] Z. Djurić,"Mechanisms of noise sources in microelectromechanical systems," *Microelectronics Reliability,* vol. 40, pp. 919–932, 2000.

[37] T. A. Roessig, R. T. Howe, and A. P. Pisano, "Nonlinear mixing in surface-micromachined tuning fork oscillators," in *Frequency Control Symposium, 1997., Proceedings of the 1997 IEEE International,* pp. 778–782, 1997.

[38] R. Boudot and E. Rubiola, "Phase noise in RF and microwave amplifiers," *IEEE Transactions on Ultrasonics Ferroelectrics & Frequency Control,* vol. 59, pp. 2613–2624, 2012.

[39] A. L. McWhorter, and L. Laboratory, "1/f noise and related surface effects in germanium," 1955.

[40] D. B. Leeson, "A simple model of feedback oscillator noise," *Proceedings of the IEEE,* vol. 54, pp. 329–330, 1966.

[41] M. Horák and V. Papež, "Very low noise DC power supply," in *International Conference on Applied Electronics,* pp. 99–102, 2012.

[42] B. Kim, R. N. Candler, M. A. Hopcroft, M. Agarwal, W. T. Park, and T. W. Kenny, "Frequency stability of wafer-scale film encapsulated silicon based MEMS resonators," *Sensors & Actuators A Physical,* vol. 136, pp. 125–131, 2007.

[43] Z. Djuric, I. Jokic, M. Franklovic, and O. Jaksic, "Influence of adsorption-desorption process on resonant frequency and noise of micro- and nanocantilevers," in *International Conference on Microelectronics,* vol. 1, pp. 243–246, 2002.

[44] M. A. Hopcroft, W. D. Nix, and T. W. Kenny, "What is the Young's Modulus of Silicon?," *Journal of Microelectromechanical Systems,* vol. 19, pp. 229–238, 2010.

[45] C. Bourgeois, E. Steinsland, N. Blanc, and N. F. De Rooij, "Design of resonators for the determination of the temperature coefficients of elastic constants of monocrystalline silicon," in *Frequency Control Symposium, 1997., Proceedings of the 1997 IEEE International,* pp. 791–799, 1997.

[46] A. Hajjam, A. Logan, and S. Pourkamali, "Doping-Induced Temperature Compensation of Thermally Actuated High-Frequency Silicon Micromechanical Resonators," *Journal of Microelectromechanical Systems,* vol. 21, pp. 681–687, 2012.

[47] T. Pensala,. Jaakkola, M. Prunnila, and J. Dekker, "Temperature compensation of silicon MEMS Resonators by Heavy Doping," in *Ultrasonics Symposium,* pp. 1952–1955, 2011.

[48] H. Tada, A. E. Kumpel, R. E. Lathrop, J. B. Slanina, P. Nieva, P. Zavracky, et al., "Thermal expansion coefficient of polycrystalline silicon and silicon dioxide thin films at high temperatures," *Journal of Applied Physics,* vol. 87, pp. 4189–4193, 2000.

[49] M. Chen and G. A. Rincon-Mora, "Accurate electrical battery model capable of predicting runtime and I-V performance," *IEEE Transactions on Energy Conversion,* vol. 21, pp. 504–511, 2006.

[50] Q. Xie, S. Yue, M. Pedram, D. Shin, and N. Chang, "Adaptive thermal management for portable system batteries by forced convection cooling," in *Design, Automation & Test in Europe Conference & Exhibition,* pp. 1225–1228, 2013.

[51] W. Guo, W. M. Healy, and M. C. Zhou, "Experimental study of the thermal impacts on wireless sensor batteries," in *IEEE International Conference on Networking, Sensing and Control,* pp. 430–435, 2013.

[52] D. Szente-Varga, G. Horvath, and M. Rencz, "Thermal characterization and modelling of lithium-based batteries at low ambient temperature," in *International Workshop on Thermal Inveatigation of ICS and Systems,* 2008. Therminic, pp. 128–131, 2008.

[53] Y. H. Park and K. C. Park, "High-fidelity modeling of MEMS resonators. Part I. Anchor loss mechanisms through substrate," *Journal of Microelectromechanical Systems,* vol. 13, pp. 238–247, 2004.

[54] R. Lifshitz and M. L. Roukes, "Thermoelastic Damping in Micro- and Nano-Mechanical Systems," *Physical Review B,* vol. 61, pp. 5600–5609, 1999.

[55] M. Bao and H. Yang, "Squeeze film air damping in MEMS," *Sensors & Actuators A Physical,* vol. 136, pp. 3–27, 2007.

[56] G. Bahl, R. Melamud, B. Kim, S. A. Chandorkar, J. C. Salvia, M. A. Hopcroft et al., "Model and Observations of Dielectric Charge in Thermally Oxidized Silicon Resonators," *Journal of Microelectromechanical Systems,* vol. 19, pp. 162–174, 2010.

[57] G. Lewicki, "Fixed charge in Cr-metallized MOS capacitors," *Journal of Applied Physics,* vol. 47, pp. 1552–1559, 1976.

[58] S. C. Vitkavage, E. A. Irene, and H. Z. Massoud, "An investigation of Si-SiO$_2$ interface charges in thermally oxidized (100), (110), (111), (511) silicon," *Journal of Applied Physics,* vol. 68, pp. 5262–5272, 1990.

[59] S. Kalicinski, M. Wevers, and I. D. Wolf, "Charging and discharging phenomena in electrostatically-driven single-crystal-silicon MEM resonators: DC bias dependence and influence on the series resonance frequency," *Microelectronics Reliability,* vol. 48, pp. 1221–1226, 2008.

Index

1/f noise 53, 63, 164, 255

A

acceleration resolution 74, 207, 248
Accelerometer 1, 25, 63, 251
additive noise 94, 139, 142, 216
Allan deviation 247
Allan variance (AVAR) 67, 89, 212
Amplificiation factor 1, 178, 188, 238
amplifier 25, 41, 176, 248
amplitude detector 70, 79, 82, 147
amplitude modulation path 136, 148
amplitude-stiffness 63, 70, 90
automatic amplitude control (AAC) 44, 66, 147, 150
axial load 5, 64, 166

B

bandwidth 29, 42, 164, 249
Beam 1, 163, 181, 247
bending moment 8, 12, 18
Bias 1, 42, 81, 252
Bias instability 49, 79, 90, 133
Bias stability 14, 65, 90, 133

Brownian noise 44, 69, 74
Buffer 76, 82, 115, 206

C

calibration 244, 246, 247, 251
capacitive sensing 19, 25, 49, 164
Charge 27, 54, 121, 241
Charge-induced drift 231, 241
charge sensing 37, 54, 77, 116
chopper frequency 79, 80, 81
chopper stabilization 79, 80, 84, 85
clamped-clamped beam 167, 170
clock frequency 103, 104, 106, 108
CMOS 29, 57, 63, 87
Comb 26, 64, 194, 195
comparator 65, 97, 111, 206
compensation 29, 89, 205, 206
continuous-time 45, 49, 76, 99
Correlated double sampling (CDS) 50
Counter 102, 122, 207, 251

D

damping 142, 150, 202, 241
dead zone 101, 112, 119, 121
deflection 8, 167, 189, 214
DEFT 167, 169
differential 28, 57, 125, 238
differentiator 45, 54, 77, 106

About the Editor

Yong Ping Xu graduated from Department of Physics, Nanjing University, Nanjing, China, in 1977 and received his PhD degree from School of Electrical Engineering, University of New South Wales, Sydney, N.S.W., Australia, in 1994.

From 1978 to 1987, he worked at Qingdao Semiconductor Research Institute in China as an IC Design Engineer, Deputy R&D Manager, and Director. From 1996 to 1998, he was a Lecturer at University of South Australia, Adelaide, Australia. Later, he joined the Department of Electrical & Computer Engineering, National University of Singapore, in 1998 where he is currently working as an Associate Professor. His current research includes integrated interface and readout circuits for MEMS and sensors and biomedical circuits and systems.

Professor Xu is a member of Technical Program Committee of Symposium on VLSI Circuits (VLSI) since 2017. He served as a member of IEEE International Solid-State Circuits Conference (ISSCC) from 2014 to 2018 and Technical Program Committee of IEEE Asian Solid-State Circuits Conference (A-SSCC) from 2009 to 2013. He was the Organizing Committee Chair for A-SSCC 2013 and has been a member of Steering Committee of A-SSCC since 2014. He also served as a TPC Co-chair of 2007 and 2009 IEEE International symposium on Radio Frequency Integration Technology (RFIT). He was the co-recipient of 2007 DAC/ISSCC Student Design Contest Award and the recipient of 2004 Excellent Teacher Award from National University of Singapore. Professor Xu is a Distinguished Lecturer of IEEE Solid-State Circuits Society from 2017 to 2018.